Safety and Security Studies

Safety and Security Studies

Editors

M. Guarascio
University of Rome "La Sapienza", Italy

C.A. Brebbia
Wessex Institute, UK

F. Garzia
University of Rome "La Sapienza", Italy

M. Lombardi
University of Rome "La Sapienza", Italy

WIT*PRESS* **Southampton, Boston**

Editors:

M. Guarascio
University of Rome "La Sapienza", Italy

F. Garzia
University of Rome "La Sapienza", Italy

C.A. Brebbia
Wessex Institute, UK

M. Lombardi
University of Rome "La Sapienza", Italy

Published by

WIT Press
Ashurst Lodge, Ashurst, Southampton, SO40 7AA, UK
Tel: 44 (0) 238 029 3223; Fax: 44 (0) 238 029 2853
E-Mail: witpress@witpress.com
http://www.witpress.com

For USA, Canada and Mexico

Computational Mechanics International Inc
25 Bridge Street, Billerica, MA 01821, USA
Tel: 978 667 5841; Fax: 978 667 7582
E-Mail: infousa@witpress.com
http://www.witpress.com

British Library Cataloguing-in-Publication Data

A Catalogue record for this book is available
from the British Library

Library of Congress Catalog Card Number: 2018932130

ISBN: 978-1-78466-313-1
eISBN: 978-1-78466-314-8

The texts of the papers in this volume were set individually by the authors or under their supervision. Only minor corrections to the text may have been carried out by the publisher.

Printed in Great Britain by Lightning Source, UK.

The material contained herein is reprinted from a special edition of the International Journal of Safety and Security Engineering, Volume 8, Number 2 published by WIT Press.

Preface

The papers contained in this volume address the most recent developments in the theoretical and practical aspects of the important fields of Safety and Security Engineering, which due to its special nature, brings together in a systematic way, many disciplines of engineering, from the traditional to the most technologically advanced.

The papers cover areas such as crisis management, security engineering, natural disasters and emergencies, terrorism, IT security, man-made hazards, risk management, control, protection and mitigation issues, and many others.

The Editors are grateful to the reviewers, as well as to all authors for their contributions.

The Editors
2018

Contents

AN EMERGENCY RESPONSE DECISION MATRIX AGAINST TERRORIST ATTACKS WITH IMPROVISED DEVICE IN CHEMICAL CLUSTERS

BEHNAZ HOSSEINNIA, NIMA KHAKZAD & GENSERIK RENIERS
Safety and Security Science Group, Delft University of Technology, The Netherlands.

ABSTRACT
Chemical industrial areas may constitute potential targets for deliberate actions by terrorists. Terrorists having sufficient knowledge of chemical process operations or plant layout may take advantage of improvised explosive devices (IEDs) to cause major events such as fire, explosion and toxic gas dispersion with cross-border consequences in chemical clusters. Thus, an efficient cluster-wise emergency plan to enhance the promptness and efficacy of responding to such attacks is crucial. In this study, the effects of blast wave caused by IEDs are assessed and its potentiality in triggering domino scenarios are analysed. A decision tree is developed to determine the emergency level of each company within the cluster based on the attack outcomes. Furthermore, an alert notification system is set based on a decision matrix. Finally, the identified emergency levels and the alert levels are presented in form of a multi-plant decision matrix. The application of the developed methodology is demonstrated in a case study.
Keywords: chemical industrial area, decision matrix, decision tree, emergency response, improvised explosive device, terrorist attack

1 INTRODUCTION

Prior to 9/11 attacks, terrorist threats had been recognized for a long time, but a successful terrorist attack on chemical plants was assumed to be quite unlikely [1]. However, the security risks has become an important concern for the chemical industry ever since [2, 3].

Chemical industrial areas due to handling large quantities of hazardous chemicals are potential targets for deliberate actions by terrorists. Compared to safety accidents in the chemical industry, security events could be much worse in terms of the extent and severity and may affect many people in nearby communities. Despite recent concerns about weapons of mass destruction (WMD), explosions are by far the most common cause of mass-casualty incidents associated with terrorism [4]. A successful attack aimed at direct or indirect damage of target equipment by bombs or fire arms can trigger domino scenarios [5] and cause damages to multiple process units or eventually several neighbouring industrial site [6, 7]. The terrorist attacks to Iraq's oil terminal in 2004 and largest refinery in 2015 are examples of terrorist attacks by improvised explosive devices (IEDs) to chemical plants [8, 9]. IEDs are unconventional kinds of bombs usually carried or delivered in a vehicle, package, by a person, or concealed on the roadside [10].

A number guideline and approaches have been proposed regarding the emergency response and preparedness for security threats in chemical industry such as [11], but there is no specific procedure available for managing terrorist attacks in multi-plant chemical industrial areas. Thus, an efficient cluster-wise emergency response seems crucial for enhancing the promptness of reacting to the attack. In this regard, the present study is aimed at introducing an innovative multi-plant response decision matrix that creates an overview of emergency levels and alert levels for individual plants within a cluster in order to help them respond in a pre-agreed procedure to terrorist attacks with IEDs. Each emergency level indicates the

DOI: 10.2495/SAFE-V8-N2-187-199

potential actions that an individual could adopt to prepare for and respond to catastrophic terrorist attacks. The alert levels can help the security decision makers plan for increasing the security of other critical assets in the chemical cluster in the prospect of imminent threats or attacks in the future.

The developed methodology is presented in the next section. The application of the methodology is illustrated in Section 3, while Section 4 concludes the paper.

2 METHODOLOGY

2.1 Overview of the methodology

The approach is based on five main components which are (i) identifying the target assets, (ii) developing and analysing the most likely attack scenarios, (iii) determining the emergency levels, (iv) developing alert notification system and (v) establishing cluster decision matrix.

2.2 Identifying the target assets

In security terms, assets for a chemical facility are defined as people (both on-site and off-site), information (trade secrets, confidential business information and etc.) and property (buildings, process equipment, control systems and etc.) [12]. Not all assets have equal value to adversaries [13]. In the first step, the assets of each company are identified, and then, they are prioritized based on their attractiveness as a target for bombing attack scenarios. Some of the relevant attractiveness factors adopted from Ref [13–15] in the present study are as follow:

- The potential to cause maximum damage (causalities, economic loss) including process equipment with significant quantities of flammable or toxic chemicals, the central control room, and utility units.
- Ease of access for adversary, for example: proximity of the assets to the facility boundary, public road, parking lot or dock area
- Recognisability of the target, etc.

2.3 Developing and analysing the most likely attack scenarios

IEDs can come in many forms, ranging from a small pipe bomb to a sophisticated device capable of causing massive damages [10]; the extent of damage caused by an IED depends on its size, construction and placement. After possible target assets are identified, a number of attack positions can be selected based on the chemical plant layout and the location of the target assets. For example, a parking lot within a short distance from storage tanks area, or a road between two chemical plants a road near control room or dock areas can be considered as potential attack areas. Table 1 shows a selection of the possible IEDs and their explosive capacity (TNT equivalent mass) based on the maximum amount of material that could reasonably fit into a container or vehicle [10].

2.3.1 Consequence analysis and domino effect considerations

In order to evaluate the effects of IEDs on structure and equipment, two parameters are considered: the weapon size, measured in equivalent kilograms of trinitrotoluene (TNT) as

Table 1: List of possible IEDs and their explosion capacity [10].

Threat description	Explosive mass (kg) (TNT equivalent)
Pipe bomb	2.3
Suitcase bomb	23
Sedan	454
Moving truck	13608
Semitrailer	27216

shown in Table 1, and whether the generated peak overpressure exceeds the threshold values needed for causing structural damage [16, 17]. Several previous publications [18–20] have provided correlations relationships for the shock wave produced by IEDs. In this study, the relationship developed by Bounds [19] is used to evaluate the stand-off distances (i.e. the distance measured from the center of gravity of the explosion to the area that the IED can cause damages). In eqn (1), P is the peak overpressure threshold (bar), r is the stand-off distance (m) and $m_{TNT,eq}$ is the equivalent TNT mass (kg) for a specific improvised device from Table 1.

$$P = \frac{m_{TNT,eq}^{\frac{1}{3}}}{r} + 4.4\frac{m_{TNT,eq}^{\frac{2}{3}}}{r^2} + 14.0\frac{m_{TNT,eq}}{r^3}. \qquad (1)$$

The calculated stand-off distances, based on the blast overpressure damage thresholds presented in Table 2, can be used to determine the process equipment being impacted by the IEDs blast waves. The potential damages at process units could result in fire, explosion, and fragment projection that may lead to further damages inside or outside the company premises by triggering domino effects. In this study, to determine which units are possibly impacted by such events, the received fire heat radiation or explosion overpressure by a nearby unit is compared with respective threshold values listed in Table 2.

Furthermore, to estimate the probability of domino effects, the damage probabilities of target units can be calculated by using Probit functions [22]. Having the Probit values, Y calculated from Table 3, the damage probability, D, can be calculated by eqn (2), where ϕ is the cumulative density function of a standard normal distribution [23].

$$D = \phi\,(Y-5). \qquad (2)$$

Furthermore, based on the calculated D, five cut-off levels for domino effects' probability are defined and categorized in five ranking levels.

Table 2: Damage thresholds due to overpressure and heat radiation for different equipment categories [21]. The same overpressure thresholds are used to determine the possibility of damage due to IED or domino effects.

Equipment category	Overpressure (bar)	Heat radiation (kW/m^2)
Atmospheric vessel	0.22	15
Pressurized vessel (toxic material)	0.2	45
Pressurized vessel (flammable material)	0.31	45

Table 3: Models for domino probability used in this study [22].

Escalation vector and primary scenario	Target equipment	Model for domino probability*
All radiation scenarios	Atmospheric	$Y=12.54 - 1.847 \ln(ttf)$; $\ln(ttf)= -1.128 \ln(I) -2.667\times 10^{-5} V+9.877$
	Pressurized	$Y=12.54 - 1.847 \ln(ttf)$; $\ln(ttf)= -0.947 \ln(I) + 8.835 V^{0.032}$
All overpressure scenarios	Atmospheric	$Y=-18.96+2.44\ln(P_s)$
	Pressurized	$Y=-42.44+4.33\ln(P_s)$

*ttf: time to failure(s), I: radiation intensity on target equipment (kW/m^2), V: equipment volume (m^3), P_s= peak overpressure on target equipment (Pa).

2.4 Determining emergency levels

A ranking criterion is provided to classify the different attack scenarios based on the attack severity and its potential impact on the company, the cluster, and the public. In the criteria table, the attack's consequence is ranked in five levels of severity, similar to the scale used in API 780 [13] from very low to very high. Table 4 provides the details for consequence ranking.

Table 4: Attack scenarios' consequence ranking.

Rank	Consequence Category	Loss of Life	Environmental Impact	Property Damage Impact	Domino effect
1	Very low	No injuries or fatalities, First aid required	No environmental impact	Limited localized minor damage	**Unlikely-** no chance of additional escalation hazards ($D < 10E\text{-}6$)
2	Low	injuries that are not widespread but only in the vicinity of the incident location	Minor environmental impacts to immediate incident site area only	Significant localized damage of some equipment/ buildings, no major repair is required.	**Moderate-** may cause additional hazards ($10E\text{-}6 \leq D < 0.001$)
.	.	.	.	.	.
.	.	.	.	.	.
.	.	.	.	.	.
5	Very High	Possibility of any offsite fatalities from large-scale toxic or flammable release; possibility of multiple onsite fatalities (inside cluster).	Major environmental impact on-site and/or offsite (e.g. large-scale toxic contamination of public waterway)	Major structural damage widespread inside cluster Extensive off-site damage	**Currently occurring-** ($0.5 \leq D \leq 1.0$)

After the attack consequences are analyzed and ranked, a decision tree is used to determine the emergency levels at the companies within the chemical cluster in case of a terrorist attack. Figure 1 shows a part of the developed decision tree if the bombing attack causes fire and explosion. Five emergency levels are defined from level 0 (Informative Alarm) to Level 4 (high-high alarm) based on the severity of the attack. Therefore, the attack's consequence is the key element in evaluating the emergency levels.

It is important not to confuse a 'security response' intended to engage and hopefully neutralize the adversaries with the broader "emergency response" that follows an attack and attempts to reduce the severity of the event and lessen the consequences in terms of loss of life and destruction of property or production capability. In this study, each emergency level indicates responsible people, response strategies and resources.

For example, if a terrorist attack impacts a number of companies within the cluster, the emergency level at each company can be levels 0 or 1 (for a company that is not/little affected and its impact is within the company's boundary), level 2 (for a companies that is moderately affected and it may impact outside the company's premises), level 3 (for a company that is highly affected and its impact reaches other companies or outside the cluster and level 4 (for a company that is severely affected and its impact causes further damage inside/outside the cluster).

2.5 Alert Level notification system for security response

When either there is the possibility of an imminent threat or an attack has happened against a particular asset, it should rapidly be communicated across the industrial area to determine appropriate security responses and to increase the protection of target assets and make it difficult for an adversary to harm or damage those assets [24]. A cluster alert notification system

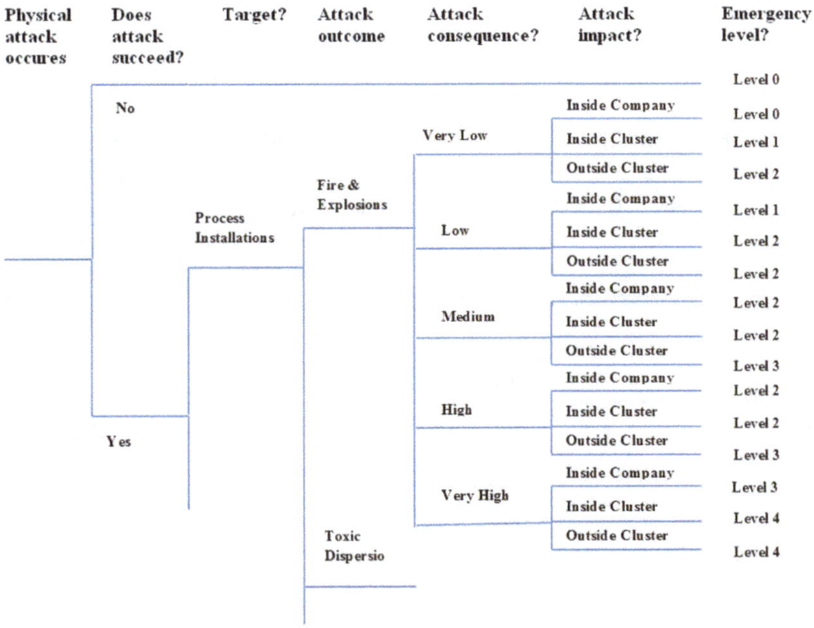

Figure 1: Emergency level decision tree.

offers help to security decision makers within the chemical cluster in order to analyze and prioritize the information regarding the potential risks at individual plants. According to API/NPRA Security Vulnerability Assessment [14] each alert level indicates what security measures need to be implemented at the facility based on the level of the threat. In this study for determining the alert levels, two parameters are considered: the emergency level of the actual attack (evaluated from the previous section) and the likelihood of terrorist attack (L) against other critical asset within the cluster.

L can be defined as the multiplication of Threat (T) and asset' Attractiveness (A) [13]. For the purpose of this study, the likelihood of IED attacks are assumed to be very highly denoted with probability of 0.8, indicating that there is a credible threat against similar assets.

Furthermore, for evaluating A, the most critical asset – from the threat' perspective – at individual companies are identified and the respective ranking level is determined from Table 5. The multiplication of the evaluated ranking probability of A to 0.8, yields to L and its respective ranking level is shown in Table 6.

A 5 × 5 matrix is developed in order to determine the alert level within the companies of the cluster. The emergency level of an attack is placed on the vertical axis while the likelihood of the attack on the horizontal axis. The matrix is presented in Fig. 2, while Table 7 provides specific definition of each alert level.

Table 5: Target asset attractiveness ranking [13].

Ranking	**Descriptor**	**Conditional Probability of the act**	**Threat interest ranking**
1	Very Low	$0.0 \leq A \leq 0.2$	Threat would have little to no level of interest in the asset.
2	Low	$0.2 < A \leq 0.4$	Threat would have some degree of interest in the asset, but it is not likely to be of interest compared to other assets.
3	Medium	$0.4 < A \leq 0.6$	Threat would have a moderate degree of interest in the asset relative to other assets.
4	High	$0.6 < A \leq 0.8$	Threat would have a high degree of interest in the asset relative to other assets.
5	Very High	$0.8 < A \leq 1.0$	Threat would have a very high degree of interest in the asset, and it is a preferred choice relative to other assets.

Table 6: Likelihood of attack ranking.

Ranking	**Descriptor**	**Likelihood of attach**
1	Very low	$0.0 \leq L \leq 0.2$
2	Low	$0.2 < L \leq 0.4$
3	Medium	$0.4 < L \leq 0.6$
4	High	$0.6 < L \leq 0.8$
5	Very high	$0.8 < L \leq 1.0$

ER Level	Likelihood of Attack				
	1	2	3	4	5
0	Alert Level 0	Alert Level 1	Alert Level 1	Alert Level 2	Alert Level 2
1	Alert Level 1	Alert Level 1	Alert Level 2	Alert Level 2	Alert Level 3
2	Alert Level 1	Alert Level 2	Alert Level 2	Alert Level 3	Alert Level 3
3	Alert Level 2	Alert Level 2	Alert Level 3	Alert Level 3	Alert Level 4
4	Alert Level 2	Alert Level 3	Alert Level 3	Alert Level 4	Alert Level 4

Figure 2: Alert level decision matrix.

Table 7: Alert level description (Adopted from the U.S. Homeland Security website [25]).

Ranking	Description	Alert level considerations
1	Low	Low risk of terrorist attack, normal security posture and conduct of business operations
2	Guarded	General risk of terrorist attack, heightened awareness advisory notice by nearby companies or the cluster security
3	Elevated	Significant risk of terrorist attack, increasing surveillance of critical locations. Coordinating emergency plans as appropriate with nearby companies
4	High	High risk of terrorist attacks, extend monitoring capability, Increase security posture Preparing to execute contingency procedures (such as evacuation site personnel) Restricting threatened facility access to essential personnel only
5	Severe	Severe risk of terrorist attacks, increasing or redirecting personnel to address critical emergency needs; Expand surveillance and response capability, Assigning emergency response personnel and pre-positioning and mobilizing specially trained teams or resources

2.6 **Multi-plant decision matrix**

The identified attack scenarios are placed on the vertical axis, and all the companies within the cluster are placed on the horizontal axis of the matrix. The emergency levels and the alert levels identified from the decision tree and the decision matrix is shown within each cell of the matrix for the companies being either affected by the attack or a likely target for similar attacks. The established multi-plant matrix model is partly depicted in Fig. 3.

	Company A		Company B		Company C	
	ER level	Alert level	ER level	Alert level	ER level	Alert level
SCEN 01	1	3	2	2	3	4
SCEN 02	3	4	1	2		
...						

Figure 3: Multi-Plant decision matrix example.

3 APPLICATION OF THE METHODOLOGY

3.1 Case study

In order to demonstrate the developed methodology, a chemical industrial area including three companies is taken into account (Fig. 4).

It is considered that terrorists had managed to access Company 3. They used a truck as a Vehicle Borne Improvised Explosive Device (VBIED) containing 13608 kg (TNT equivalent) explosive mass (Table 1); they have the truck parked at attack position 1 (AP1 in Fig. 1), near storage tank area 1 and dock 1. The impact of the explosion is large enough (eqn 1) to cause damage to nearby process equipment leading to a major fire. For consequence assessment, wind direction of southwest (SW), wind speed of 7 m/s, stability class D, and ambient temperature of 20°C were assumed.

3.2 Results and discussions

The potential impact radius of the explosion against the atmospheric storage tanks in area 1 (at Company 3) is calculated as 192 meter, based on the overpressure escalation thresholds of 0.22 bar. The explosion stand-off distance contour is shown in Fig. 5, while Table 8

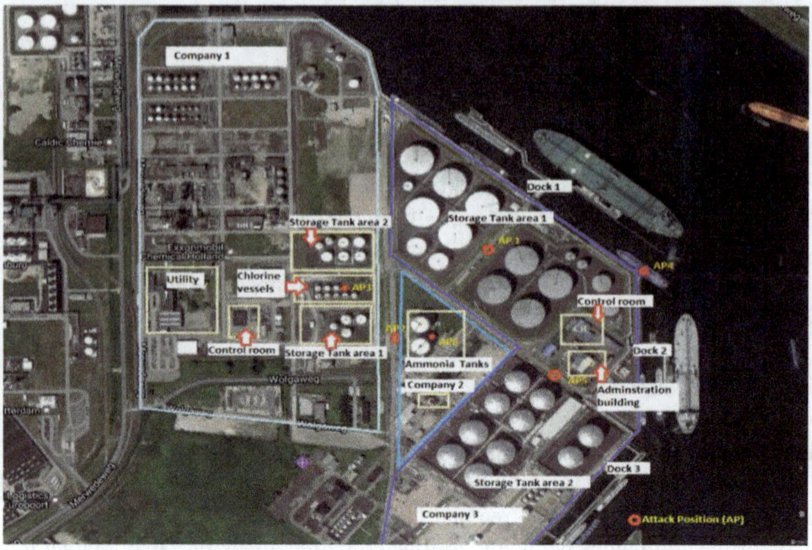

Figure 4: Chemical cluster comprising three chemical plants.

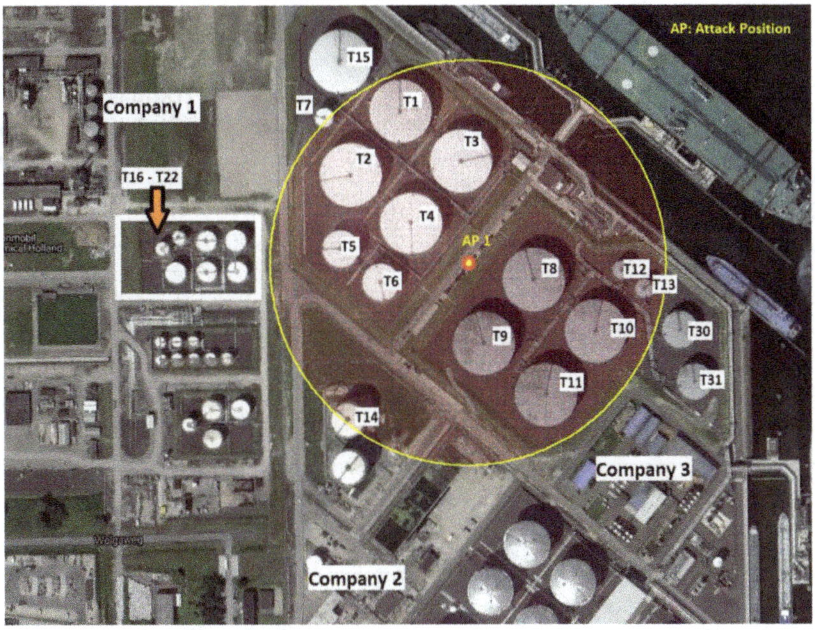

Figure 5: The impact zone (0.22 bar) of a VBIED detonation at attack position 1

Table 8: Information related to the equipment that are influenced by VBIED detonation at attack point 1.

Vessel ID	Vessel Type	Diameter (m)	Height (m)	Stored Substance	Inventory (m^3)
T1-T4	Atmospheric	60	21.2	Kerosene	54000
T5-T6	Atmospheric	35	18	Kerosene	15.586
T7	Atmospheric	18	15.2	Kerosene	3.481
T8-T11	Atmospheric	60	21.9	Benzene	55800
T12-T13	Atmospheric	18	15.2	Benzene	1,934
T14	Atmospheric	30	21.24	Ammonia	13500

summarizes the associated equipment items, which receive explosion overpressure higher than or equal to the correspondent threshold value. As indicated in Fig. 6, the attack not only has severe impact inside Company 3, it also affects the Northern part of Company 2. It is assumed that the damaged units by blast overpressure in Company 3 will initiate pool fires that are able to impact other companies within the cluster. The consequence of the pool fire scenarios are calculated (using ALOHA software) on the nearby units exposed to high heat radiation levels (greater than or equal to the threshold values in Table 2). The primary pool fire scenarios has the potential to cause further damage and trigger secondary scenarios (such as pool fire, tank fire, explosions, etc.) on target equipment. For example, the pool fire at tank T2 will affect several equipment (T16-T22) in Company 1. The results of the primary pool fire scenarios and the domino probabilities of affected units are reported in Table 9.

Table 9: Primary and secondary scenarios triggered by VBIED detonation at attack point 1 (AP1).

Primary event	Escalation Vector	Affected Units	Primary Scenario	Escalation vector	Secondary units	Damage Probability	Possible Secondary Scenarios
VBIED Blast at AP1	Over-pressure	T1–T7	Pool fire	Radiation	T15, T16–T22	0.022 7.35E-7	Pool fire, Tank fire,
	Over-pressure	T8–T13	Pool fire	Radiation	T30, T31	0.005	Pool fire
	Over-pressure	T14	Leakage/ toxic release	-	-		-

Besides the damage to the properties and equipment, there would be causalities not only at Company 3 but also the other companies within the cluster. The off-site (outside cluster) causalities may happen due to indirect effects of the attack. For example the blast overpressure and projections may cause a major leakage at Ammonia storage tank (T14). The personnel of passing ships or boats in the nearby waterway may be exposed to high amounts of the toxic gas (within the AEGL-2) and would experience serious or irreversible adverse health effects. Also the gas cloud (AEGL-1) is likely to reach the residential area outside the cluster (10 km), causing notable discomfort, irritation and reversible effects to the public.

Using Table 2 and the decision tree in Fig. 1, the emergency levels are determined as follow:

Company 1 is not directly impacted by the attack, however the results from Table 9 indicates that several units in this plant are exposed to high levels of heat radiation and the domino effect is likely. The consequence ranking is medium, and it will not cause further adverse impacts outside the company premises. Therefore, the emergency level can be identified as 2.

Company 2 is directly influenced by the blast overpressure, and T14 (Ammonia storage tank) is damaged and may cause huge impact both inside and outside the cluster. Therefore, the emergency level is 3.

A large part of Company 3 is damaged by the attack, there is a severe environmental and property damage, and domino effect is almost certain within the company. The consequence is very high, and Dock 1 area and the nearby waterway are impacted accordingly. Therefore, the emergency level is 4.

Since the threat (T) for this security event is considered very high for each chemical plant with the probability 0.8, and the most attractive target assets at the three companies are identified, the likelihood of attack can be calculated as in Table 10.

After the emergency levels and the likelihood of attack for each company is evaluated, the alert levels can be predicted using the decision matrix in Fig. 3. The final result is presented in the multi-plant decision matrix in Fig. 6.

Table 10: Likelihood of attack at the chemical cluster.

Com-pany	Most critical Target Asset from threat's perspective	Attractiveness (A) probability	Threat (T) probability	Likelihood of attack (L=A×T)	Likelihood of attack ranking
1	Chlorine pressurized vessels	1.0	0.8	0.8	5- Very High
2	Ammonia Tank	0.8	0.8	0.64	4- High
3	Tank area 1 & 2	0.9	0.8	0.72	4- High

	Company 1		Company 2		Company 3	
	ER Level	Alert Level	ER Level	Alert Level	ER level	Alert level
Attack at the car park area in Company 3	2	3	4	3	4	4

Figure 6: The multi-plant decision matrix.

4 CONCLUSIONS

Chemical industrial areas can be critical targets to terrorist attacks with improvised explosive devices. Such attacks may affect several neighbouring facilities, the whole cluster and even the nearby communities. In the present study, in order to help emergency responders and security decision makers within the chemical cluster an efficient cluster-wise decision tool, comprising a decision tree and decision matrices, was developed that determines the emergency and alert levels at each company. Each identified emergency level identifies the potential actions that an individual could adopt to prepare for and respond to a terrorist attack scenario. Likewise, each alert level indicates to what extent to increase the security of other critical assets due to either the possibility of an imminent threat or the occurrence of an actual attack against a particular asset.

ACKNOWLEDGEMENTS

The authors acknowledge the financial support provided by Leiden-Delft-Erasmus Centre for Safety and Security (LDE-CSS), The Netherlands.

REFERENCES

[1] Baybutt, P., Strategies for protecting process plants against terrorism, sabotage and other criminal acts. *Homeland Defence Journal*, **2**, pp. 1–7, 2003.

[2] Belke, J.C., Chemical accident risks in US industry—A preliminary analysis of accident risk data from US hazardous chemical facilities. *Loss Prevention and Safety Promotion in the Process Industries,* **2001**, pp. 1275–1314, 2000. https://doi.org/10.1016/B978-044450699-3/50041-0

[3] Bajpai, S. & Gupta, J.P., Site security for chemical process industries. *Journal of Loss Prevention in the Process Industries*, **18**(4), pp. 301–309, 2005. https://doi.org/10.1016/j.jlp.2005.06.011

[4] Arnold, J.L., Tsai, M.C., Halpern, P., Smithline, H., Stok, E. & Ersoy, G., Mass-casualty, terrorist bombings: epidemiological outcomes, resource utilization, and time course of emergency needs (Part I). *Prehospital and Disaster Medicine*, **18**(03), pp. 220–234, 2003. https://doi.org/10.1017/s1049023x00001096

[5] Landucci, G., Reniers, G., Cozzani, V. & Salzano, E., Vulnerability of industrial facilities to attacks with improvised explosive devices aimed at triggering domino scenarios. *Reliability Engineering & System Safety*, **143**, pp. 53–62, 2015. https://doi.org/10.1016/j.ress.2015.03.004

[6] Pavlova, Y. & Reniers, G., A sequential-move game for enhancing safety and security cooperation within chemical clusters. *Journal of Hazardous Materials*, **186**(1), pp. 401–406, 2011. https://doi.org/10.1016/j.jhazmat.2010.11.013

[7] Reniers, G. & Soudan, K., A game-theoretical approach for reciprocal security-related prevention investment decisions. *Reliability Engineering & System Safety*, **95**(1), pp. 1–9, 2010. https://doi.org/10.1016/j.ress.2009.07.001

[8] *Suicide bombers in boat attack on Iraq oil terminal*, available at: https://www.theguardian.com/world/2004/apr/25/iraq2

[9] *ISIS launches attacks at Iraq's largest oil refinery,* available at: http://english.alarabiya.net/en/News/middle-east/2015/04/12/ISIS-militants-attack-Iraq-s-largest-oil-refinery.html

[10] Kennett, M., Letvin, E., Chipley, M. & Ryan, T., Federal Emergency Management Agency (FEMA), US Department of Homeland Security, United States of America, US Deptartment of veterans affairs and United States of America, risk assessment: a how-to guide to mitigate potential terrorist attacks against buildings. *FEMA Risk Management Series*, 2005.

[11] Bajpai, S. & Gupta, J.P., Terror-proofing chemical process industries. *Process Safety and Environmental Protection*, **85**(6), pp. 559–565, 2007. https://doi.org/10.1205/psep06046

[12] *Site security guidelines for the U.S. chemical industry*, 2001, http://www.ohlhausen.com/wp-content/uploads/2013/07/chemical-site-guidelines.pdf

[13] API Recommended Practice 780, *Security risk assessment methodology for the petroleum and petrochemical industries, document draft.* Washington, DC: American Petroleum Institute, 2012.

[14] API/NPRA, *Security vulnerability assessment*. Washington, DC: American Petroleum Institute Publishing Services, 2003.

[15] Argenti, F., Landucci, G., Spadoni, G. & Cozzani, V., The assessment of the attractiveness of process facilities to terrorist attacks. *Safety Science*, *77*, pp. 169–181, 2015. https://doi.org/10.1016/j.ssci.2015.02.013

[16] Salzano, E. & Cozzani, V., The analysis of domino accidents triggered by vapor cloud explosions. *Reliability Engineering & System Safety*, **90**(2), pp. 271–284, 2005. https://doi.org/10.1016/j.ress.2004.11.012

[17] Cozzani, V., Gubinelli, G. & Salzano, E., Escalation thresholds in the assessment of domino accidental events. *Journal of Hazardous Materials*, **129**(1), pp. 1–21, 2006. https://doi.org/10.1016/j.jhazmat.2005.08.012

[18] Bounds, W.L., *Design of blast resistant buildings in petrochemical facilities*. ASCE Publications, Reston, VA, 1997.

[19] Alonso, F.D., Ferradas, E.G., Minarro, M.D., Aznar, A.M., Gimeno, J.R. & Pérez, J.F.S., Consequence analysis by means of characteristic curves to determine the damage to buildings from the detonation of explosive substances as a function of TNT equivalence. *Journal of Loss Prevention in the Process Industries*, **21**(1), pp. 74–81, 2008. https://doi.org/10.1016/j.jlp.2007.08.002

[20] Price, M.A. & Ghee, A.H., Modeling for detonation and energy release from peroxides and non-ideal improvised explosives. *Central European Journal of Energetic Materials*, **6**(3–4), pp. 239–254, 2009.

[21] Cozzani, V., Antonioni, G., Landucci, G., Tugnoli, A., Bonvicini, S. & Spadoni, G., Quantitative assessment of domino and NaTech scenarios in complex industrial areas. *Journal of Loss Prevention in the Process Industries*, **28**, pp. 10–22, 2014. https://doi.org/10.1016/j.jlp.2013.07.009

[22] Cozzani, V., Gubinelli, G., Antonioni, G., Spadoni, G. & Zanelli, S., The assessment of risk caused by domino effect in quantitative area risk analysis. *Journal of Hazardous Materials*, **127**(1), pp. 14–30, 2005. https://doi.org/10.1016/j.jhazmat.2005.07.003

[23] Khakzad, N., Khan, F., Amyotte, P. & Cozzani, V., Domino effect analysis using Bayesian networks. *Risk Analysis*, **33**(2), pp. 292–306, 2013. https://doi.org/10.1111/j.1539-6924.2012.01854.x

[24] Sullivant, J., *Building a corporate culture of security: strategies for strengthening organizational resiliency*. Butterworth-Heinemann, 2016.

[25] *National terrorism advisory system*, available at: https://www.dhs.gov/national-terrorism-advisory-system

AN ACTIVE MAN-IN-THE-MIDDLE ATTACK ON BLUETOOTH SMART DEVICES

TAL MELAMED
Tech Leader, AppSec Labs, Israel.

ABSTRACT

In the last years, the Internet of Things (IoT) has become integral part of our lives and its influence is expected to exponentially increase in the next years. For several reasons, however, the development of IoT has not gone hand in hand with an adequate reinforcement and consolidation of our security and privacy, despite the serious impact that IoT vulnerabilities may have on our digital and physical security. Bluetooth Low Energy (BLE), also known as Bluetooth Smart, is the most popular protocol for interfacing smart devices, wearables, and medical equipment. This contribution surveys the key security issues in the BLE protocol and discusses a possible architecture for BLE Man-in-the-Middle (MitM) attacks together with the related necessary equipment. In addition, after introducing some of the available tools for hacking BLE, a case-study based on their use was presented, which describes a MitM attack between a Bluetooth smart device and its designated mobile app. The case-study well exemplifies how easily, given the required proximity to the target, a possible hacker can control the data and, in some instances, even the mobile device itself, when connecting it to a BLE device.

Keywords: BLE security; Bluetooth Low Energy; Bluetooth Smart; IoT security

1 INTRODUCTION

Bluetooth technology was first introduced as 'Wibree' by Nokia in 2006. With the adoption of the Bluetooth Core Specification Version 4.0 in 2010, Bluetooth Low Energy (BLE) became the main Bluetooth standard, known under the names of 'BLE' and 'Bluetooth Smart' [1]. BLE was designed as a smaller, cheaper, and power-efficient technology aimed at supporting the increasing growth in the use of IoT devices, which is excepted to reach the number of 30.7 billion devices in 2020 [2]. Due to its low cost and ease of implementation, about one third of the existing smart devices use BLE. Bluetooth is broadly used with IoT devices/applications and smart products in the healthcare, fitness, beacons [3], security, and home-entertainment industries, and nowadays, we encounter BLE in almost every aspect of our lives (e.g. in wearables, sensors, medical devices, security products, etc.). Natively supported by the most popular Mobile operation systems (including iOS, Android, Windows Phone and BlackBerry), as well as by macOS, Linux, and Windows 8/10, by 2018, Bluetooth Smart is expected to be supported out-of-the-box by more than 90% of Bluetooth-enabled smartphones [4].

By underling some of the security vulnerabilities of the BLE, this paper aims to raise concern about the crucial need to devise new strategies for developing adequate IoT security protocols that will be able to keep up with the increasing diffusion of IoT devises and with the growth in the use of BLE. In my contribution, I will first examine some of the fundamental security issues in the BLE protocol mainly relying on Mike Ryan's survey presented in 2013 [5]. I will then discuss in details a possible architecture and the related equipment required to perform MitM attacks under BLE. After presenting two available tools that use this architecture, I will offer a case-study which exemplifies on a practical level the most important issues discussed in the paper. Based on the above-mentioned tools that use the MitM architecture, the chosen case-study describes a practical MitM attack attempted on a mobile application, which communicates with a BLE-based sport bracelet.

DOI: 10.2495/SAFE-V8-N2-200-211

2 KEY SECURITY ISSUES IN THE BLE PROTOCOL

BLE protocol has proved insecure against eavesdropping [5]. As intercepting and decrypting the BLE communication is relatively simple, potential hackers can easily perform Man-in-the-Middle (MitM) attacks, i.e. attacks in which the attacker manages to place himself in the middle of the communication between the smart device and the designated mobile app.

In the following paragraphs, I briefly survey some key concepts of Bluetooth Smart together with their impact on the global security level of the protocol. Further information about BLE encryption, pairing, key-generation, signing data and other security-related issues are described in the BLE Security specification [6].

2.1 Security manager

The Security manager is part of the Core System Architecture (Fig. 1) and is in charge of the security capabilities of the protocol: pairing integrity, authentication and encryption. BLE responds to MitM attacks with AES-CCM encryption. While AES encryption is considered secure, the way the encryption keys are exchanged is insecure and introduces some severe security vulnerabilities, which can be exploited by potential hackers to decrypt the data. In fact, to exchange the encryption key, Bluetooth Smart uses a custom key-exchange protocol, in which the devices exchange a temporary key (TK) and use it to create a short-term key (STK), which will be then used to encrypt the connection. The level of security of such process depends on the pairing method used to exchange the TK, for which see the following section.

2.2 Pairing

The pairing process is performed by two BLE – or BLE-supported – devices to establish keys in order to encrypt the communication. The keys can be also used to encrypt the communication in future reconnections, to verify received signed-data, or to perform random

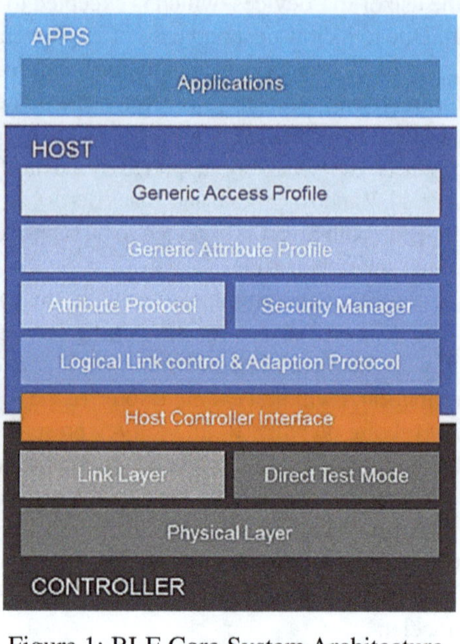

Figure 1: BLE Core System Architecture.

address resolution. BLE uses a three-phase pairing process. (i.) The initiating device sends a pairing request and, consequently, the devices exchange pairing capabilities over an insecure channel. (ii.) The devices exchange the TK and verify the use of the same TK, which is then used to generate a STK. Rather than TK and STK, newer devices (BLE 4.2) support LE Secure Connections, which use a long-term key (LTK) exchanged using Elliptic Curve Diffie-Hellman (ECDH) public key cryptography [7], a significantly stronger methodology than the standard BLE key exchange protocol. (iii.) The created key is then distributed between the devices over a secured connection and can be used to encrypt the data.

The TK is determined according to the pairing method. In version 4.0 and 4.1 of the Core Specification, BLE uses the Secure Simple Pairing model (known as LE Legacy), in which devices can choose to operate with one of the following methods, according to their specific I/O capabilities: (i.) Just Works™, (ii.) Passkey entry, and (iii.) Out-of-Bound (OOB), outlined below (2.2.1, 2.2.2, 2.2.3). It ought to be emphasized that mobile apps cannot control the pairing method, which is instead determined on the OS level.

2.2.1 Just Works™

In Just Works™, the TK is set to 0, thus offering no protection at all against MitM attacks. Despite its high level of insecurity, this is the most commonly used method today among the BLE devices. Furthermore, this is the only possible method that can be employed with devices lacking a display.

2.2.2 Passkey

In Passkey, the TK is a six-digit number combination, which the user actively inserts into one of the devices, thus manually exchanging it between the devices. For instance, if one of the devices generates a random six-digit number and shows it on its LCD display, the user will need to read the number and enter it into the other device using a keypad. Since it was proven that also this method can be bypassed [5], it is therefore recommended to follow either the OOB or Numeric Comparison pairing methods (the latter supported only by BLE 4.2).

2.2.3 Out-of-Band

In this method, the TK is exchanged using a different medium, such as Near Field Communication (NFC). The security level of OOB is determined by the security capabilities of the technology used to exchange the key. If the OOB channel is protected from MitM attacks, the BLE connection shall be also regarded as sufficiently protected. In other words, the pairing process will result immune to eavesdropping, if the OOB channel is immune. OOB pairing is the most secure method available in the BLE 4.0/4.1 protocol. Unfortunately, this method is still very uncommon: having tested dozens of smart devices, I have seen no implementation of OOB yet.

2.3 GATT

The Generic Attribute Profile (GATT) is used by BLE devices to communicate with each other using a standardized data schema (Fig. 2). The GATT describes the device roles and general behaviors employing hierarchy of services, characteristics, and attributes. According to Bluetooth Specifications, "a service is composed of characteristics or references to other services. A characteristic consists of a type (represented by a UUID), a value, a set of properties indicating the operations the characteristic supports and a set of permissions relating to security. It may also include one or more descriptors metadata or configuration flags relating to the

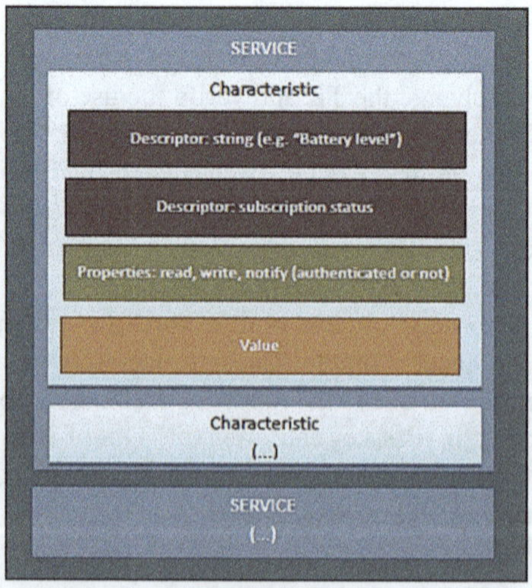

Figure 2: GATT Profile Hierarchy.

owning characteristic" [8]. Provided the proximity to a smart device, its GATT schema and generic data can be read by any BLE-supported app.

Each service contains properties that can be used by the connected app (certain properties require authentication). There are three supported events: read, write, and notify. While read and write requests transmit a single value, a notify event allows the mobile app (central) to subscribe for a certain characteristic and receive data asynchronously from the smart device (peripheral). All the communications are performed using integer-value handlers associated with the requested characteristics.

2.4 BLE connection flow

In a typical BLE connection flow (Fig. 3), a smart device continuously broadcasts an advertisement, in order to allow a mobile app to identify and connect to it. Once the mobile app is connected to the smart device, the BLE device advertises its GATT schema, providing the connected app with the necessary information.

This flow enables potential hackers to discover the device services, clone them and re-publish them, in order to lure the victim into their 'trap' (see section 3). Figure 4 illustrates in three steps a generic GATT-reader app (also referred to as BLE Scanners) reading the available services and the generic data published by the smart device (in this specific case, a 'Power Watch+' smart device).

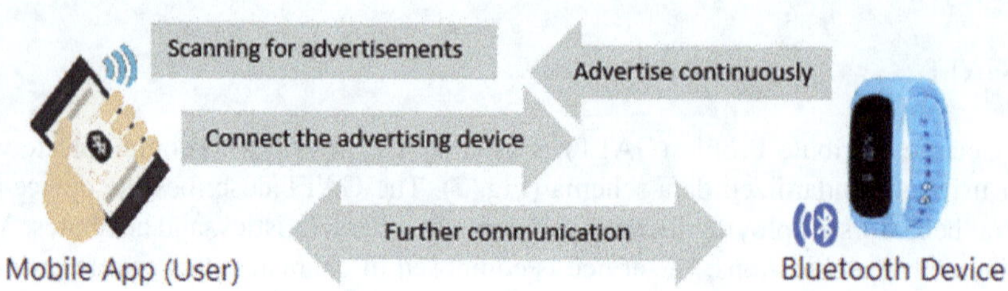

Figure 3: Typical BLE Connection Flow.

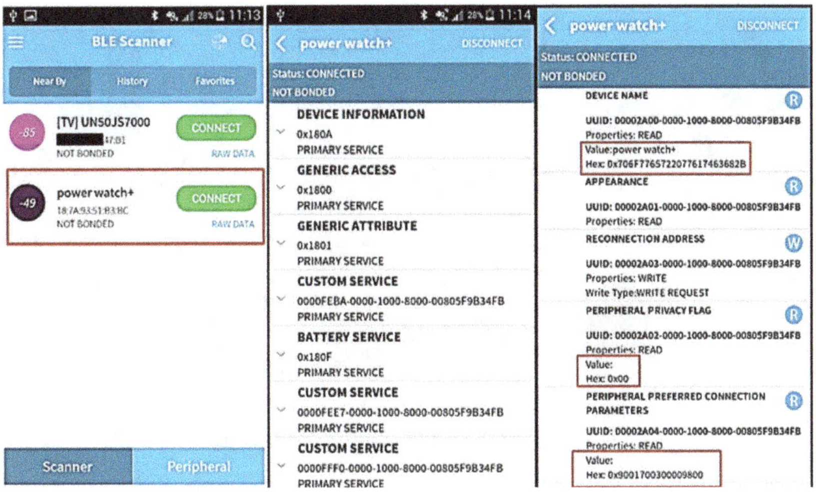

Figure 4: BLE Scanner app accessing data published by the 'Power Watch+' smart device.

3 PRACTICAL MITM ATTACK FOR BLE

A common MitM attack, in which the attacker 'sits' between two parties connecting to both ends, does not apply for Bluetooth Smart. As shown in Fig. 5, in the common MitM architecture, when one party (the client) sends the data, the MitM attacker acts as the receiving end (the server) and, vice versa, when the server sends the response back, the attacker acts as the client. This does not apply to BLE, due to a typical Bluetooth technology limitation for which the device can be connected to only one party and, therefore, cannot connect to both ends simultaneously.

3.1 BLE MitM architecture

In MitM attacks for BLE, a common MitM cannot connect simultaneously to both ends (smart device; mobile app). Therefore, a BLE MitM needs to make use of two BLE

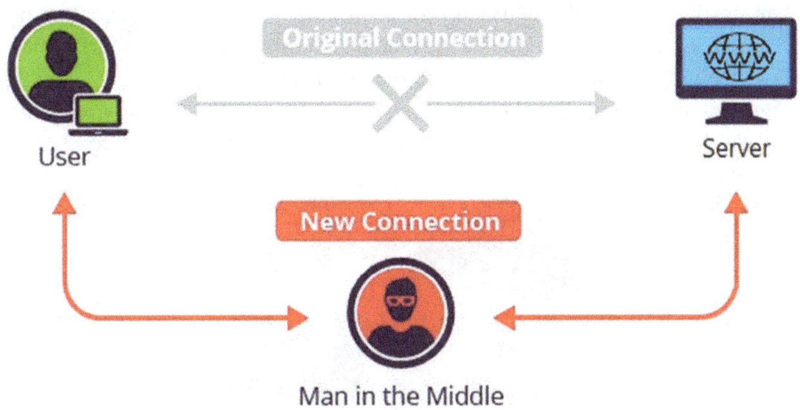

Figure 5: Common MitM architecture.

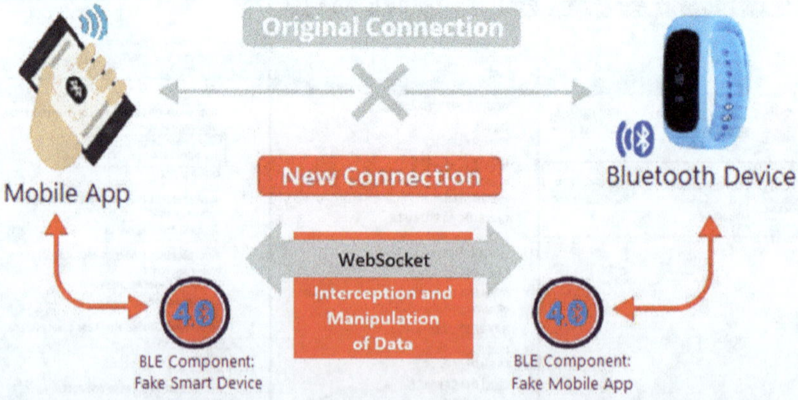

Figure 6: Practical man-in-the-middle architecture for BLE.

components capable of acting together: one connects to the mobile app acting as the smart device, while the other connects to the smart device acting as the mobile app. Once the smart device is connected to one of the components (fake mobile app), it can no longer publish itself or enable any other application to connect to it. At this point, it is required a different protocol to enable the two components – the fake mobile app and the fake smart device – to exchange the data that each of them received from, respectively, the smart device and the mobile app. A useful protocol to complete the MitM attack is WebSocket [9], which enables a two-way communication between two hosts, one running code in a controlled environment and the other opting-in to communications from that code. As shown in Fig. 6, by using the discussed architecture, the MitM attack can succeed, thus allowing the hacker to intercept and even modify the data sent over the BLE channel.

3.2 Necessary equipment

To support the described MitM architecture for BLE, the attacker requires two BLE 4.0 components. The most popular component used for this purpose is the CSR 8510-based USB dongle [10], which supports BLE (Fig. 7).

By connecting both dongles to the computer – or, better, to two different machines – the attacker can easily perform the MitM attack according to the architecture outlined above (Section 3.1; Fig. 6). Specifically, one of the dongles (acting as the fake smart device) is used to connect to the mobile app, while the other (acting as the fake app) to the smart device. In order to intercept and manipulate the data transferred between the mobile app and the smart device, the two dongles communicate with each other using a WebSocket code.

Figure 7: Typical CSR 4.0 adapters.

4 AVAILABLE TOOLS BASED ON THE BLE MITM ARCHITECTURE

Although there are other possible MitM architectures for BLE, the one described in this paper (Section 3) has the advantage that has been already tested and proved to be stable on both Linux and Windows operations systems. In what follows, I introduce two available tools on which the above-mentioned architecture is implemented: GATTacker (Section 4.1) and BtleJuice (Section 4.2). Both tools produce excellent results enabling a MitM attack for BLE and, among other many advantages, support two important features: (i.) Hooking, which enables hackers to write a code that can be executed upon a BLE event (read, write or notify), and Replay attack, which enables hackers to re-transmit an event that was previously sent [11]. Both (and other) tools can be accessed from my GitHub page (available at https://github.com/nu11p0inter). The case-study described in Section 5 uses both the above-mentioned tools and features to hack the designated mobile app.

4.1 GATTacker

GATTacker is a Node.js package for BLE MitM developed by Slawomir Jasek (Securing) and presented at the Black Hat USA conference in 2016 [12]. After installing the prerequisite libraries (i.e. noble, a NodeJS BLE central module; bleno, a Node.js module for implementing BLE peripherals), GATTacker can scan and copy BLE advertisements and services which can then be used to run a cloned (fake) version of the smart device (peripheral). Once the fake smart device is provided with the IP address of the machine to which the dongle acting as fake mobile app (central) is connected, the WebSocket connection is established and functions as an active MitM, which enables the hacker to intercept and manipulate the transmitted data.

4.2 BtleJuice

BtleJuice is a framework to perform MitM attacks on Bluetooth Smart devices, which was developed by Damien Cauquil (Digital Security) and first presented at the DefCon 24 conference in 2016 [13]. BtleJuice includes two components, which require to be run on independent machines in order to operate two Bluetooth 4.0+ adapters simultaneously. Also in this case, the two components communicate with each other over WebSocket protocol. BtleJuice includes a web interface and – among other useful features – presents Replay GATT operations (Replay attack) and On-the-fly data modification capabilities (Hooking).

5 CASE-STUDY: HACKING A BLUETOOTH SMART MOBILE APP

In what follows, I attempt to demonstrate on an empirical base the notions surveyed so far, by describing in details and step-by-step a practical MitM attack which I performed on a BLE device for research purposes. The BLE device chosen for the MitM attack and described in the case-study is a smart sport bracelet (Dax-Hub SW-28 Smart Bracelet) designed to monitor sport activities and health-related data (e.g. calories, distance, etc.). The data detected by the smart bracelet are synchronized with the designated app, called PowerSensor, which can be downloaded from the App Store and Goole Play. The case-study shows that, giving the proximity to the device, an attacker can easily intercept and manipulate the data transmitted between the mobile app and the BLE device, operation that can result in the modification of the data and in a partial mobile takeover from remote (limited to the BLE protocol range, which is ~100m).

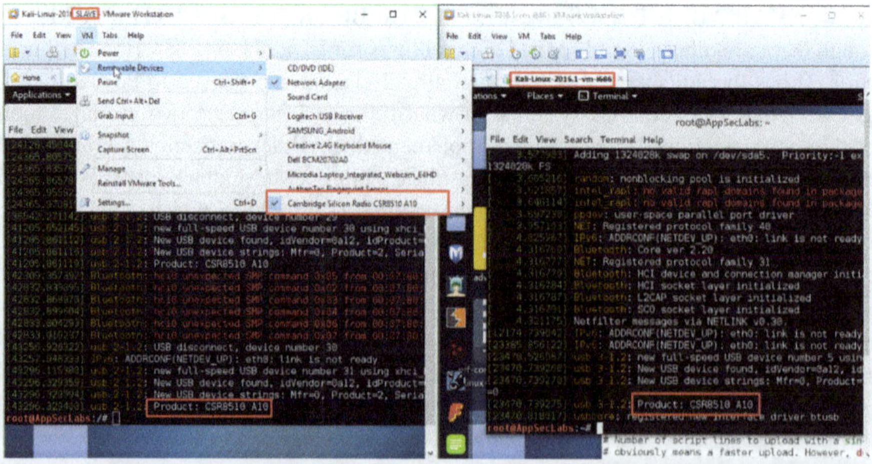

Figure 8: Each CSR dongle is connected to a different VM.

5.1 Preparing the environment

Before carrying out the planned MitM attack, I downloaded a Kali Linux virtual image (available at https://kali.org) and installed all the required libraries and tools (see Section 4). The next step was that of inserting the CSR 4.0 dongle (Section 3.2) and connecting it to the Virtual Machine (VM). Since most tools require two different machines to run the BLE components, I decided to clone the VM, making sure that each dongle was connected to a different VM, as shown in Fig. 8.

5.2 Data Interception via MitM

After the preliminary procedure described above (Section 5.1), I began the actual MitM attack by running the GATTacker central (ws-slave) on one VM and by running the GATTacker peripheral configured to the ws-slave's IP address. In order to find the advertisements published

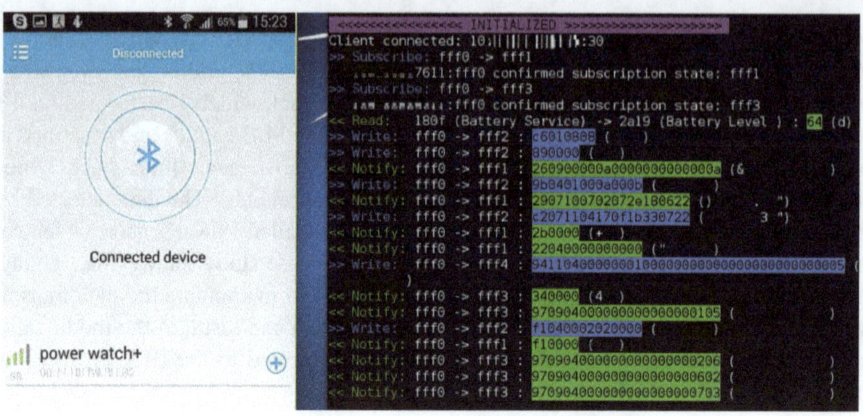

Figure 9: MitM attack based on the use of GATTacker.

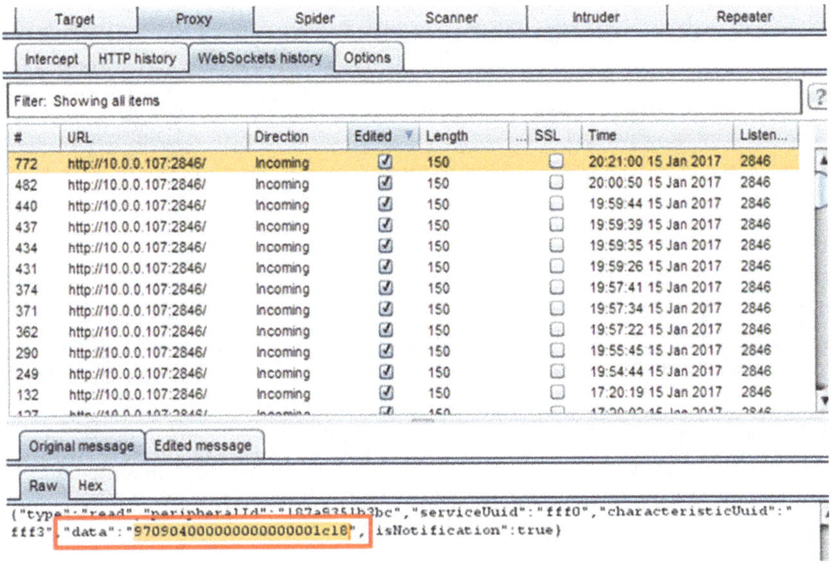

Figure 10: The BLE communication data transmitted via Burp.

by all the nearby BLE devices, it is necessary to run the scan command. However, to carry out the MitM attack, it is mandatory to detect and copy also the services of the targeted smart device. Therefore, to achieve this result, I ran the scan command specifying the MAC address of the BLE sport bracelet. This operation enables GATTacker to copy and store the discovered advertisements and services on the machine, where they will be identified by their specific MAC address. Once I obtained the advertisements and services of the BLE bracelet, I used them to simulate a fake device. In fact, by advertising the name of the targeted smart device (in this case, 'power watch+'), GATTacker can lure the victim to connect to the fake device and,

Figure 11: Hooking notification event using GATTacker.

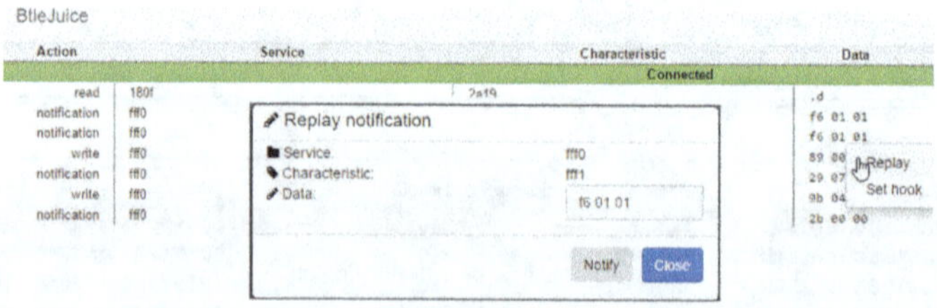

Figure 12: Replay Attack feature in BtleJuice.

when the mobile app scans for a nearby device (in the case-study PowerSensor mobile app) it locates the fake advertisement. At this point, the mobile app was successfully connected to the bracelet through MitM, thus allowing me to view all the communication data, which in this specific case were represented by long hexadecimal values, as shown in Fig. 9. In general, it ought to be noted that, if the victim has previously paired the mobile app with the smart device, the mobile app will automatically connect to the fake device, provided that the latter presents the MAC address of the real smart device using MAC spoofing [14].

Considering that setting a proxy in the middle of the WebSocket connection of the BLE components can provide additional capabilities (e.g. on-the-fly data modifications, brute-force, replay attacks, etc.), I set the peripheral component to go through Burp (i.e. a proxy tool that supports WebSocket communication, available at https://portswigger.net/burp/) rather than directly to the central component. This step allowed me to intercept and modify the data using Burp Intercept feature, as shown in Fig. 10.

5.3 Data manipulation via notification hooking

Unfortunately, Burp does not support hooking, providing only the on-the-fly ability, which can modify only one intercepted request at a time. Therefore, during the simulated MitM attack for the case-study, I had to make use also of the hooking feature of GATTacker which allows to execute custom code upon a received event (i.e. read, write, notify). The analysis of the data exchanged between the smart bracelet and the PowerSensor app during the simulated attack has shown that when a sport exercise activity is launched in the mobile app (e.g. jumping rope, treadmill, etc.), the app subscribes to a notification event – identified as fff3 (see Fig. 9) – in which the bracelet monitors the exercise activity and sends the data to the PowerSensor app in real-time. The GATTacker hooking feature enabled me to add a hook to the notification event identifier and, consequently, to execute code capable of automatically modifying the data on every request sent from the bracelet to the mobile app (see, Fig. 11). As a result of the hook, the values sent from the bracelet to the PowerSensor – which at the time was used for a treadmill activity – were modified to an exaggerated kilometrage. Obviously, modifying exercise data poses no actual threat to the targeted victim. Yet, while I deliberately chose to simulate in the case-study a harmless attack as proof of capabilities, in a real-world scenario hackers can target different life-depending devices, such as diabetic/ blood-sugar monitors, as well as security devices, such as electronic locks and alarms.

5.4 Mobile camera hijacking via replay attack

Besides the data monitoring features mentioned above, the smart bracelet chosen for the case-study supports the use of the mobile camera and music system directly from the wristband. Therefore, using BtleJuice, I was able to identify such requests, after invoking the functionality from the bracelet. As mentioned in Section 4, BtleJuice includes Replay Attack capabilities, which allowed me to replay a request previously transmitted between the bracelet and the mobile app. After initializing the MitM attack using BtleJuice, the web interface showed a notification on the fff0 service that the data f60101 was used by the smart bracelet to take a picture on the victim's mobile device. Using the Replay function, I was then able to take pictures with the mobile camera directly from my own computer at any given time (Fig. 12). With the same method, I was able also to play music directly on the victim's mobile (for obvious reasons, I cannot supply concrete proof in the paper that I actually succeeded to take pictures and play music as a result of the MitM attack).

6 CONCLUSIVE NOTES

The present paper confirms that BLE is insecure and vulnerable against passive eavesdropping, as demonstrated already in 2013 [5]. In particular, I have shown (Sections 3 and 4) that a passive eavesdropping can easily become an active MitM attack that enables a possible hacker not only to listen to the communication, but also to intercept and manipulate the data. Furthermore, with the case study presented in Section 5, I have demonstrated that, in some instances, by performing a MitM attack, hackers can even control from remote the mobile device used to communicate with the Bluetooth smart device.

It ought to be noted that with the release of the Bluetooth Core Specification version 4.2, BLE Security has been significantly improved by the new LE Secure Connections pairing model, which includes the Elliptical Curve Diffie-Hellman (ECDH) algorithm for key exchange and, in addition to the three standard pairing models (outlined in §2.2), the numeric comparison method. Similarly, additional security and privacy-related features are added in the Bluetooth Core Specification v5 [8], recently released by Bluetooth.

Despite these important improvements in BLE Security, it is vital to be aware and fully understand the limitations of the smart devices that we use rather than blindly relying on them. In particular, it is of essential importance to implement security protections on the application-side to guarantee users a sufficient protection against malicious activity. Ensuring full protection against complex hacking attacks may prove difficult and perhaps utopic at the present stage. Yet, a valuable step in this direction can be achieved by implementing additional and updated security controls, such as data encryption and signature, strong authentication and authorization mechanisms, and other security best practices.

ACKNOWLEDGEMENTS

The study presented in this paper is part of a broader research on IoT security carried out at AppSec Labs, a leading Israeli application security consulting company specialized in mobile and IoT security (https://appsec-labs.com/iot-security). A preliminary version of this study was originally presented at OWASP Israel January 2017 [15].

REFERENCES

[1] Bluetooth, S.I.G, SIG introduces bluetooth low energy wireless technology, the next generation of Bluetooth wireless technology, *press release*, 2009.

[2] Columbus, L., Roundup of internet of things forecasts and market estimates. *Forbes*, 2015.

[3] Dudhane, N.A. & Pitambare, S.T., Location based and contextual services using bluetooth beacons: new way to enhance customer experience. *Lecture Notes on Information Theory,* **3**(1), 2015.

[4] Janiak, S., *Three ways bluetooth smart technology enables innovation for the internet of things*, available at: https://blog.bluetooth.com/three-ways-bluetooth-smart-technology-enables-innovation-for-the-internet-of-things. (accessed 30 April, 2017).
[5] Ryan, M., Bluetooth: with low energy comes low security. *WOOT*, 2013, available at: https://www.usenix.org/system/files/conference/woot13/woot13-ryan.pdf. (accessed 30 April, 2017).
[6] Bluetooth S.I.G., *Proprietary information security, bluetooth low energy*, available at: https://www.bluetooth.com/~/media/files/specification/bluetooth-low-energy-security.ashx. (accessed 30 April, 2017).
[7] Lauter, K., The advantages of elliptic curve cryptography for wireless security. *IEEE Wireless Communications,* **11**(1), pp. 62–67, 2004. https://doi.org/10.1109/mwc.2004.1269719
[8] Bluetooth Specifications, available at: https://www.bluetooth.com/specifications. (accessed 30 April 2017).
[9] Fette, I. & Melnikov, A., The WebSocket Protocol. *Internet Engineering Task Force*, RFC 6455, 2011, available at: https://tools.ietf.org/html/rfc6455. (accessed 30 April 2017).
[10] CSR8510 Specification, available at: http://www.csr.com/sites/default/files/csr8510.pdf. (accessed 30 April 2017).
[11] Syverson, P., A taxonomy of replay attacks, computer security foundations workshop VII, CSFW 7, *IEEE*, pp. 187–191, 1994.
[12] Jasek, S., *GATTacking Bluetooth Smart devices*, BlackHat USA, 2016, available at: http://gattack.io/whitepaper.pdf. (accessed 30 April 2017).
[13] Cauquil D., *BtleJuice: the Bluetooth Smart MitM Framework, DEF CON 24 Internet of Things Village*, 2016, available at: https://www.youtube.com/watch?v=lcn07TclnS0. (accessed 30 April 2017).
[14] Cardenas, E.D., Mac Spoofing–an introduction. *GIAC Security Essentials Certification*, 2013.
[15] Melamed, T., R U aBLE? BLE Application Hacking. *OWASP*, 2017, available at: http://sl.owasp.org/melamed17. (accessed 30 April 2017).

A GAME ORIENTED APPROACH TO MINIMIZING CYBERSECURITY RISK

SCOTT MUSMAN & ANDREW J. TURNER
The MITRE Corporation, McLean, Va, USA.

ABSTRACT
Information and Communication Technology (ICT) systems are now ubiquitous in all aspects of our society. With an ability to create ICT incident effects via cyberspace, criminals can steal information or extort money, terrorists can disrupt society or cause loss of life, and the effectiveness of a military can be degraded. These threats have caused an imperative to maximize a system's cyber security resilience. Protecting systems that rely on ICT from cyber-attacks or reducing the impacts that cyber incidents cause is a topic of major importance. In this paper, we describe an approach to minimizing cybersecurity risks called Cyber Security Game (CSG), where CSG can be viewed as a form of model-based system security engineering. CSG is a method and supporting software that quantitatively identifies mission outcome focused cybersecurity risks and uses this metric to determine the optimal employment of security methods to use for any given investment level. CSG maximizes a system's ability to operate in today's contested cyber environment by minimizing its mission risk. The risk score is calculated by using a cyber mission impact assessment (CMIA) model to compute the consequences of cyber incidents, and by applying a threat model to a system topology model and defender model to estimate how likely attacks are to succeed. CSG takes into account the widespread interconnectedness of cyber systems, where defenders must defend all multi-step attack paths and an attacker only needs one to succeed. It employs a game theoretic solution using a game formulation that identifies defense strategies to minimize the maximum cyber risk (MiniMax), employing the defense methods defined in the defender model. This paper describes the approach and the models that CSG uses.
Keywords: cybersecurity, game theory, return on investment, risk assessment, risk management

1 INTRODUCTION

Information and communications technology (ICT) is now integral to almost every aspect of our daily activities. The detrimental aspect of introducing ICT dependencies, however, is that this makes us susceptible to impacts from cyber incidents. Criminals can steal and extort money or information, terrorists can disrupt society or cause loss of life, and the effectiveness of a military can be degraded, all because of an ability to create incident effects in cyberspace, and often without any requirement of physical proximity. Protecting ICT from cyber incident effects or reducing their impacts on operational activities has become of international importance. There is an escalating imperative to identify and minimize operational cyber risk.

In almost all circumstances, we are interested in achieving operational resilience: the ability for mission systems to continue to acceptably achieve their function in the face of incidents, while also doing what's best to fulfill other security requirements. Achieving such resilience almost always must be pursued in the face of resource limitations. Often we need to justify the costs and resources needed, so a problem with trying to achieve resilience is that we need a way to measure it, including how to identify whether efforts to improve it are successful or not. If operational resilience is defined in a qualitative rather than quantitative way very little prescriptive advice can be offered to increase it. Many of today's risk management methods provide only generic guidance, or recommended best practices, and produce only risk rankings. As described by [1] risk rankings lack the information required to support optimal allocation of resources to manage those risks, and more importantly: ranking doesn't contain information of how an attacker will modify their behavior in the face of defender actions.

DOI: 10.2495/SAFE-V8-N2-212-222

The difficulty with trying to manage one's cyber risk is that it requires a holistic view of how a system fulfils its intended purpose. This requires a knowledge of the purpose, the entire range of hazards that are possible, and a need to consider all attack avenues. Due to the inter-connectedness of cyber systems attackers can exploit seemingly-non-critical cyber components to bypass security controls and other defenses. Examples include Stuxnet [2] and the Target data breach [3], where non-critical systems were compromised to reach the important ones. In other words, effective cybersecurity does not just involve defending high impact resources, but all inter-connected resources that provide a pathway to those that are mission-critical. Some have proposed techniques where a defender considers only the resources that directly cause impacts [4, 5]. But without an explicit representation of how the ICT components interconnect this approach fails to defend the seemingly non-critical resources that act as stepping stones and consequently fails to defend indirect attack paths an attacker might use to bypass defenses. An attacker can choose any attack method, ranging from those with a high expected value (payoff), to speculative probes that might reveal a weakness not immediately apparent to a defender obsessed only with "the crown jewels." To defend only against the most obvious attack paths simply allows an attacker to select the next most promising one, and so on, until some exploitable chink in the armor is discovered. Thus, without a holistic understanding of the system, one is likely to make the mistake of addressing only some of the hazards but not others, or to over-invest in addressing some hazards even though the risks from other hazards are higher. A comprehensive assessment is needed to be able to provide prescriptive advice to identify which defensive tools and methods are needed and where.

Cyber attacks and the concomitant defensive actions can be viewed as game playing between two players [6]. One way this can be represented is as a zero-sum game (one where both attacker and defender assign the same value to gain or loss). In two player, finite, zero-sum games, the game theoretic solution concepts of Nash Equilibrium and Minimax produce the same solution – where the defender tries to minimize the maximum payoff for the attacker. When implementing a technique to use to solve this type of game, we need to be concerned with our ability to define the "game state" (i.e. what game board looks like), how attacker and defender moves change the game state, and to be able to evaluate each game state. Since each mission system is going to be different, and since operational objectives are whatever they happen to be, for cyber, this means the game board must be built for each system, and we need to be able to evaluate how the modeled system can fulfill its purpose in the face of cyber incidents. This is made possible by reasoning about cyber incident effects, rather than the incident instances that cause the effects [7], and by leveraging our past work on Cyber Mission Impact Assessment (CMIA) [8, 9] as a way of assessing the goodness of a "game state." Combining the "mission impact" measure with a system topology based [10, 11] threat likelihood model allows us to estimate mission risk.

We describe a program that allows us to play a cyber security game with an objective to minimize a mission system's cyber risk. We call our software CSG, the Cyber Security Game. Our approach is theoretically grounded in game theory, yet focuses on being practically useful. We describe the specifics of how the game is formulated, and each of the models of the system needed to play the game. A cyber mission impact assessment (CMIA) model is used to identify the consequences of cyber incidents. A model is used to describe the ICT topology, connectivity, access relationships, and asset type information. Models define the defensive measures that can be employed, and a default attacker model is provided. Each of these models is algorithmically linked to aspects of the risk calculation so that changes to each can directly alter the mission risk metric. CSG is a model-based solution, meaning that user effort is focused on developing the models of the system and its usage scenario, and CSG's algorithms then operate on the models to compute mission risk and optimize the employment of security measures. The outcome of running CSG allows us to identify the optimal set of security tools and resilience techniques to reduce the mission systems' cyber risk for any given defender cost.

The next section describes our game, describes how we represent cyber mission systems, how we assess cyber threats, how we have implemented the game, and how we use it to perform a portfolio analysis of tools and methods that can reduce cyber risk. We then discuss aspects of our game program and how it solves the problem we have posed.

2 THE CYBER SECURITY GAME

CSG is an algorithmic method that relies on models that describe the system, its purpose, the threat environment, and the defender capabilities. It runs algorithms on those models to produce results. CSG algorithms automate several expert level capabilities, such as the combinatorics of possible incidents, attack path discovery, and portfolio analysis, so that analysts do not have do them manually. When aspects of the system, defenses, or threats change (i.e., new vulnerabilities are discovered), a defender merely updates the appropriate model and reruns CSG to assess the impact of the changes. An overview of CSG is provided here.

CSG's game formulation searches through the combinatorics of cyber incidents, attack paths, impacts, and defender method employments using Minimax. CSG is formulated as a two-player zero-sum game. As such it implements rational decision making, where both players work to best counteract each other's moves. CSG assumes that the attacker knows about the system they are attacking. This is reasonable because CSG's focus is on defense employment for the long term, assuming that attackers will be able to learn about and may pursue targets over the lifetime of a system. Methods for trying to deceive, delay, or deter an attacker over the short term would motivate a different game formulation. Examples of different formulations are shown in [6].

The system metric in CSG is a system risk score. This score comes from computing the different impacts that an attacker can cause combined conditioned on how difficult the system architecture and defenses makes it to cause those impacts (described later in this paper). In the game, the defender tries to minimize this value. A state in the game represents a configuration of the system, either with no defenses employed, or some status quo version of the system that one wants to improve. To compute the risk score for that state, for the attacker player generates an attack tree that looks multiple attack steps ahead to identify possible impacts. The leaf nodes of the attack tree provide an expected value (EV) computed from the impacts and the probabilities (as shown in Fig. 5). The defender then employs defenses to reduce expected value of the impacts the attacker can cause, and the game proceeds in typical game tree fashion. Starting from the initial state it employs MiniMax, to assess move pairs of a defender action (min), and the attackers revised attack tree given the defender changes (max). The resultant game tree assesses how each defender method can best reduce the risk score. Since the employment of each defender method incurs a cost, the game is over when either the game identifies the optimal set of defense methods to use when the defender has spent the amount of money they have allocated, or when a complete portfolio analysis is performed to compute the Pareto frontier for each price point.

Figure 1 demonstrates the output of CSG, where the cost and performance of each portfolio option is plotted. The red dots show the Pareto optimal portfolios. In this example, 55,296 defense portfolio options were considered. With no defenses the risk score was 8492934. With all the security methods applied, the risk was reduced to 2038408, for a cost of $250k. However, spending ~$39k, the risk can be reduced to 2779440. This represents 89% of the risk reduction for only 16% of the cost.

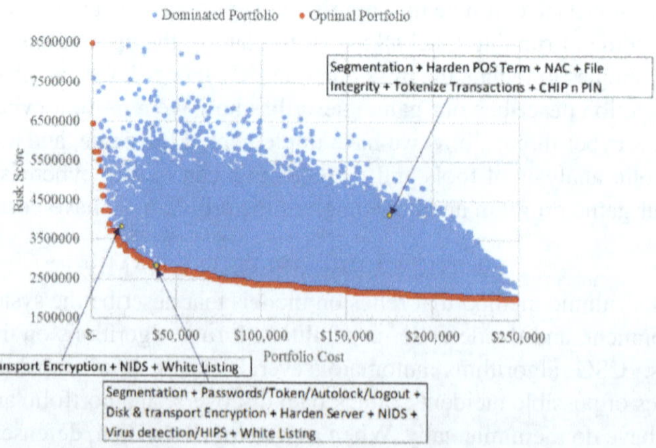

Figure 1: Security Portfolio Effectiveness vs. Cost.

CSG requires four system models. A CMIA process model computes impacts. A system topology model describes the component interconnectivity, type and trust relationships. It contains a built-in (default) attacker model. Lastly, models of the defender methods are needed. Each of these models is described below.

2.1 Modeling Incident Impacts

CSG uses CMIA process models to determine the consequence (losses) incurred from cyber incidents, and the CMIA software is embedded in CSG. CMIA makes it possible to capture mission details as an executable simulation. It models mission activities, ICT activities, activity durations, activity dependencies, ICT resources, temporal constraints, data, and control flows. ICT resources in the process model can be affected by cyber incident effects represented by the DIMFUI cyber incident effect taxonomy described in [7] and relates the occurrence of an effect to mission outcomes in the form of mission impacts. Figure 2 shows

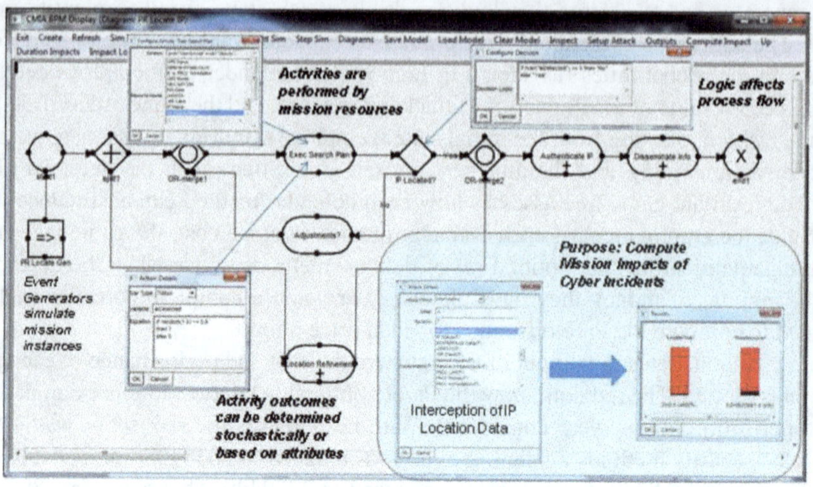

Figure 2: Process Model Showing Mission Activity Order and Control Flow Dependencies.

the CMIA tool displaying a Business Process Modeling Notation (BPMN) diagram, and shows some of the underlying details that turn the model into an executable simulation. It is outside the scope of this paper to describe CMIA in detail but for more information on the CMIA method and tool see [9, 12].

CMIA process models can be probabilistic and stochastic, allowing one to bound the uncertainties associated with the model. Running combinatorics on the set of cyber incident effects allows the criticality of cyber resources to be estimated [12] or validated by mission subject matter experts. CMIA also supports the assessment of multiple (simultaneous) incidents, allowing CSG to run through combinations of the possible cyber incident effects that will be assessed using MiniMax.

2.2 Modeling the Attacker

CSG, provides a default attacker model that defines the probability that attacks will succeed given the topological constraints that the system imparts on the attacker. For the attacker to affect ICT resources that can cause impacts, the attacker must find a pathway to access them. The attack model conditions the probability of an attack succeeding with the following characteristics:

- Whether the attacker is trying to compromise a component they can directly connect to, or whether it requires crossing a network trust boundary to access
- Whether that component is the same type as one of the components that the attacker has already demonstrated they can compromise
- Whether a component is known to be vulnerable to known exploits
- Whether the component is a server that contains one or more network services
- Whether user roles, who have access to each resource, can leverage those roles to access other components in the network to create impact

Figure 3 shows the probabilities of attacker success given topology relationships. If they first try and attack host Win 7-2 before attacking something else, the diagram shows that is it possible to attack the host from the outside with probability of success $P(S \mid OIC)$. It also shows there is a probability in becoming an inside user or subverting an insider account, where the number of accounts N, will affect the risk $N \times P(S \mid IA)$. Once a host is compromised an attacker can compromise other resources on the devices with $P(S \mid HA)$.

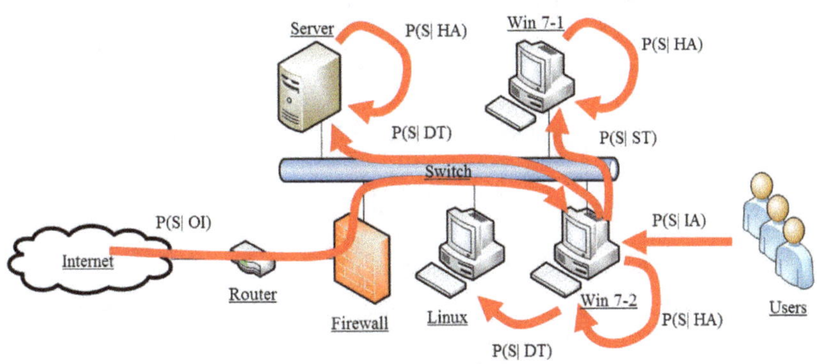

Figure 3: Default Attack Model.

Once in, the attacker can try to attack other cyber resources of the same type as one of the components already compromised then the probability is P(S|STC). If it's of a different type, then the attacker needs a new zero-day attack and has the probability $P(S \mid DTC)$. Throughout, whether the host is a client or a server will affect the probabilities. The probability of successfully navigating the network to compromise a cyber resource is computed using the chain rule shown in Equation 1. This attack model captures the basic security properties of segmentation, diversity, and least privilege and can be composed across multiple networks, trust, and segmentation boundaries. It mathematically credits a defender reducing the amount of access to components that can cause significant impacts, for diversifying components that can cause significant impacts when attacked in combination, it credits a defender for making it harder to reach the components that cause significant impacts

$$P(A_1, A_2, A_3, \ldots, A_n) = P(A_1 \mid A_2, A_3, \ldots, A_n)\,P(A_2, \mid A_3, \ldots, A_n)\,..P(A_{n-1}, \mid A_n)\,P(A_n).$$

Equation 1: General Form of the Chain Rule

2.3 Modeling System Topology and Applying the Attack Model

CSG computes impacts and then uses the attacker model to estimate the probability that the impacts will occur given the constraints of the system topology. The topology model includes cyber components, applications, data, user account groups, and access controls that enforce trust relationships. Model elements include single ICT resources as well as ICT resource pools that represent functionally identical groupings of resources of the same type. The model also requires resource type information for each ICT resource. This makes it possible to know when the same attacker exploit from an earlier step might be reused, and when it can't. The existence of connections, firewall rules, and the access of user roles define connectivity capabilities and restrictions between ICT resources. An example of a topology model is shown in Fig. 4. User group populations are also included in the model, but the diagram does

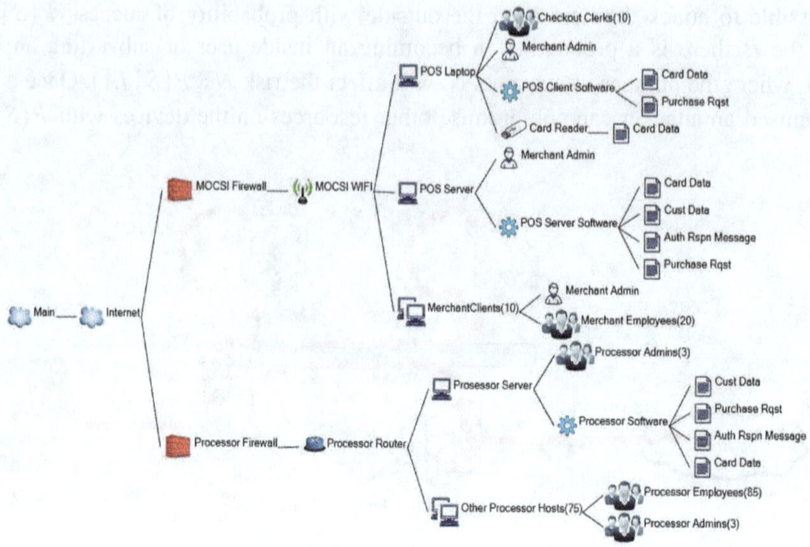

Figure 4: A Simple Topology Model for a Point of Sale System.

not show the trust relationships of who can communicate with who, that are also captured in the model.

2.4 Computing the Risk Score

The topology model makes it possible to compute attack trees. This process is illustrated in Fig. 5. The figure illustrates the complexity of analyzing even a simple four host system. All the mission impacts in the figure would be computed from a CMIA model of the system. The impacts shown include impacts from when multiple components are compromised. Even though the model in the figure contains only two subnets, the model represents a trust relationship from S1 and S2 to S4. The attack tree is then generated, using the attacker model to estimate the probability of attack steps succeeding.

In Fig. 5 only the attack steps from the internet are part of the final tree. This is possible for two reasons. First, rational attackers will always use the optimal path to compromise a component in the system. Second, we are evaluating risks in the context of MiniMax. This allows suboptimal paths to be pruned from the tree. Even so, this small four node example produces a more expansive attack tree than we can fully illustrate in the figure (only some pathways are shown). At the end of each path in the attack tree an expected value is computed (based on the worst-case – the max in MiniMax). To do this it is necessary to consider different pathways to achieve the same impacts. This is necessary because trust relationships (i.e., as imposed by firewall rules) are not necessarily directionally symmetric, so each pathway has a different probability of success. The sum of EV's at the end of each branch in the attack tree represents the risk score for the system. An alternative risk score for risk averse organizations (such as critical infrastructure components) could be the maximum EV that exists in the tree. In the next section we will discuss how defensive measures that reduce these EV's are modeled.

Because an attacker may have to compromise multiple ICT resources to cause significant impacts there is a need to look enough steps ahead to identify those cases. In CSG this number of attack steps to explore is a parameter that can be set. Usually, unless a system is

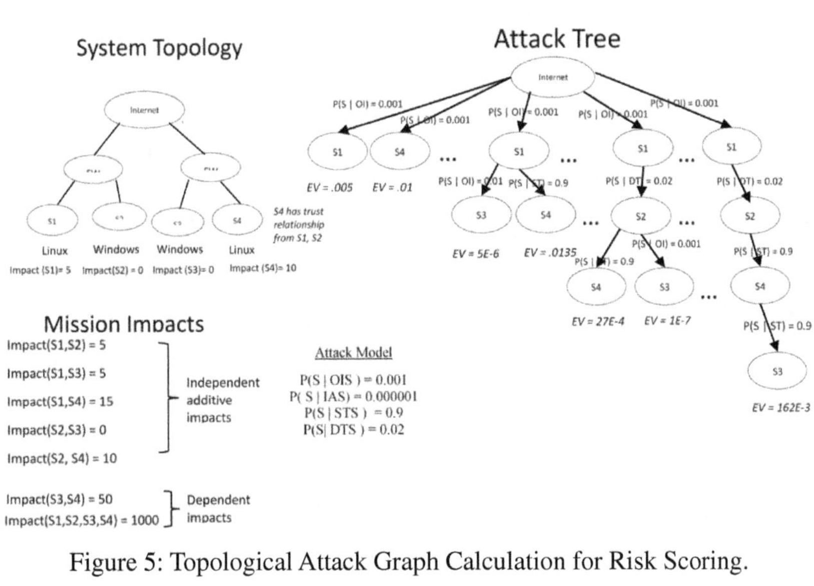

Figure 5: Topological Attack Graph Calculation for Risk Scoring.

composed of particularly layered defensive boundaries, looking 3-4 steps ahead should be sufficient.

2.5 Modeling Defender Methods

To assess defender choices CSG requires models of the defensive measures. These are shown in Fig 6. Some of the defense measures reduce the likelihood that attacks will succeed. This is typically accomplished in one of two ways. One way applies protections to the cyber resources, and the other is to change access. Each defense measure requires an assessment of how well the method is expected to work. Measures that reduce the chance of attack success are shown in the table in Fig. 6. The table lists: the measure name, cost estimates for employing the measure, and measure effectiveness assessment against each attack effect. This effectiveness assessment is defined as a score from 0 to 100. A score of 0 means that the measure is of no use to prevent the effect occurring, and a score of 100 means that the measure completely stops all the attacks that would cause that effect. An interpretation for this score is that a value of 40 should imply that 40% of the attacker exploits that we expect or anticipate as possible would no longer succeed.

Other defender measures can be represented by changes to the CMIA process model. The bottom left of Fig. 6 shows how a redundant server can be represented in the process model. Including a redundant path in the process illustrates the presence of a redundant server. If an incident only affects the original server then no impact occurs. However, since CSG looks multiple attacker steps ahead it will eventually identify the attack scenario that causes incidents on both servers.

Other defensive measures can be represented in the topology model. This is accomplished by copying the topology model and modifying it to reflect any proposed changes. The likelihood that attacks will succeed can be reduced by changing the network topology or changing

Methods that Reduce the Chance of Attack Success

Category	Method	Applies-to	Tool Name	Install Cost	Maintenance Cost	Operational Cost	Total Cost	Interruption	Modification	Fabrication	Unauthorized-Use	Interception
Encryption	Encrypt Disk & transport Layer	Mobile Dvc	LUKS & TLS	$1000	$0	$50	$1050	0	40	40	40	60
Server Configuration Management	Harden	EHR Server	Puppet	$2000	$2000	$100	$4100	20	20	20	20	20

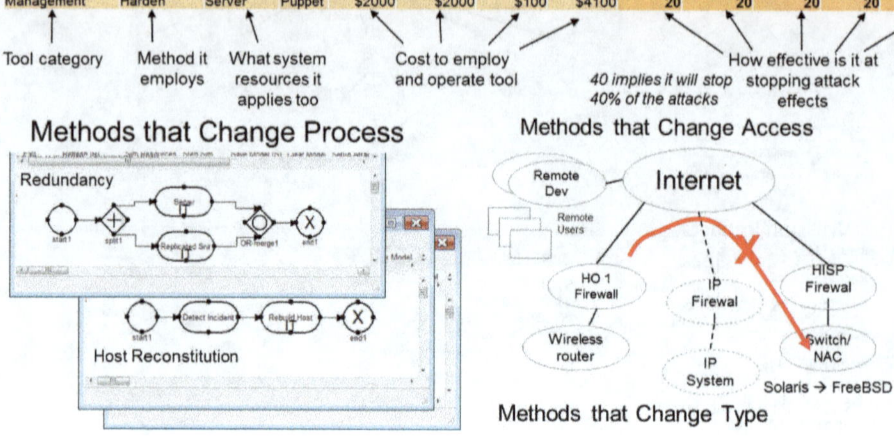

Figure 6: Modeling Defender Methods.

access controls. Risks can also be reduced by the principle of least privilege. This involves reducing the number of pooled resources (i.e., client workstations and/or users) with access. Lastly, the risks can be reduced by diversifying the components in the system, forcing the attacker to require multiple different attacks to create an impact [11].

3 SUMMARY

In this paper, we have described the Cyber Security Game (CSG). CSG is both a method and software that implements the method. CSG implements an approach that represents a quantitative, assessment of a mission system's cyber risk. CSG formalizes the information gathering activities that occur during a traditional risk assessment into computable models, artifacts that describe the system (i.e., a system topology model and impact model). CSG also algorithmically encapsulates expert capabilities in software that leverages these computable artifacts to implement a cyber risk assessment that is a consistent and comprehensive process. CSG artifacts also make it practical to keep risk assessments up to date as the system or missions change. A defender need only update the appropriate CSG model(s) to reflect changes, and rerun CSG to update the assessment.

CSG explores the possible attack paths for as-is or as-may-be system topologies. It uses MiniMax search to investigate how attackers react to defender actions. Finally, it performs a portfolio analysis to identify optimal risk reduction portfolios. Such computations are challenging for non-experts to perform consistently, especially without assisting tools. Implementing these in a tool form can help "level the playing field" for organizations that perform risk assessments, by reducing the need for risk assessment specialists and by producing artifacts that clearly capture the system, mission, threat and defense and assumptions that the assessment used.

When remediating risks it is easy to over invest in defending one portion of the system at the expense of underinvesting in defending others. CSG combats this by looking across the entire set of incidents, and outcomes. It apportions defenses to reduce the risks systematically. Each step in the game attempts to mitigate the largest risks existing at that point in the game. At the same time, it looks some steps ahead to avoid making a locally greedy decision. Each defender move may cause attacking some other component(s) to have the largest payoff for an attacker. Therefore, CSG can answer important cybersecurity questions like, "how much diversity is enough?" CSG answers this by being able to diversify components until the best attacker move becomes one that attacks other components not being protected by diversity. At that point, in a game theoretic sense, additional diversity is wasted given that the attacker already has a higher payoff elsewhere. By exploring the game-tree, CSG portfolios are always balanced to reduce overall risk.

When used with a decision target (i.e., reduce the risk by 70%) or a cost threshold (i.e., how much money one wants to spend), CSG can be used to determine whether the available set of defense methods can achieve some risk reduction target and identify the optimal set of defense methods needed to reach that target. If there is no a priori decision target, CSG can also be used to perform an entire portfolio analysis. This makes it possible to identify the Pareto frontier allowing the user to identify the optimal set of defense methods to use at each given price point, and to understand if and when there is a diminishing return on their investment.

THIS PAPER HAS BEEN APPROVED FOR PUBLIC RELEASE; DISTRIBUTION UNLIMITED. CASE NUMBERS 15-3140 AND 16-3240

REFERENCES

[1] Cox, A., Some limitations of risk = threat × vulnerability × consequence for risk analysis of terrorist attacks. *Risk Analysis*, **28**(6), pp. 1749–1761, 2008. https://doi.org/10.1111/j.1539-6924.2008.01142.x

[2] Lagner, R., *To Kill a Centrifuge*. The Langner Group, 2013.

[3] United States Senate Committee on Commerce, Science, and Transportation, *A "Kill Chain" Analysis of the 2013 Target Data Breach*. United States Senate , Washington, D.C., USA, 2014.

[4] Garvey, P.R. & Patel, S.H., *Analytical frameworks to assess the effectiveness and economic-returns of cybersecurity investments*. Military Communications Conference (MILCOM), 2014 IEEE, Baltimore, MD, USA, 2014.

[5] Carin, L., Cybenko, G. & Hughes, J., Cybersecurity strategies: the QuERIES methodology. *Computer,* **41**(8), pp. 20–26, 2008. https://doi.org/10.1109/mc.2008.295

[6] Roy, S., Ellis, C., Shiva, S., Dasgupta, D., Shandilya, V. & Wu, Q., *A Survey of Game Theory as Applied to Network Security*. 43rd Hawaii International Conference on System Sciences (HICSS), Koloa, HI, USA, 2010.

[7] Temin, A. & Musman, S., *A Language for Capturing Cyber Impact Effects. MITRE Technical Report MTR-10344*. MITRE Corporation, Washington DC, 2010.

[8] Musman, S., Temin, A., Tanner, M., Fox, F. & Pridemore, B., *Evaluating the Impact of Cyber Attacks on Complex Missions*. 5th International Conference on Information Warfare and Security, Dayton, OH, USA, 2010.

[9] Musman, S.T.A., *A Cyber Mission Impact Assessment Tool*. Homeland Security Technologies Conference, Boston, MA, 2015.

[10] Jajodia, N.S., Topological vulnerability analysis. *Cyber Situational Awareness, Advances in Information Security,* 2010.

[11] Wang, L., Jajodia, S. & Noel, S., *k-zero day safety: Measuring the security risk of networks against unknown attacks*. European Symposium on Research in Computer Security, Athens, Greece, 2010.

[12] Musman, S., Tanner, M., Elsaesser, E. & Lewis, L., *A Systems Engineering Approach to Crown Jewels Estimation and Mission Assurance Decision Making*, Proceedings of the IEEE Symposium on Computational Intelligence in Cyber Security, Paris, France, 2011.

[13] Musman, S., Tanner, M., Temin, A. & Elsaesser, E., *Computing the Impact of Cyber Attacks on Missions*. 2011 IEEE International Systems Conference (SysCon), Montreal, QC, 2011.

[14] Musman, S. & Agbolosu-Amison, S., *A Measurable Definition of Resiliency Using "Mission Risk" as a Metric*, MITRE Corp, McLean, VA, USA, 2014.

[15] Dhanjani, N., Rios, B. & Hardin, B., *Hacking: The Next Generation*, O'Reiley Media Inc., Sebastopol, CA, 2009.

[16] Noel, S., Ludwig, J., Jain, P., Jhonson, D., Thomas, R., McFarland, F., King, B., Webster, S. & Tello, B., *Analyzing Mission Impacts of Cyber Actions (AMICA)*. ATO IST-128 Workshop on Cyber Attack Detection, Forensics and Attribution for Assessment of Mission Impact, Istanbul, Turkey, 2015.

[17] Nguyen, N., Alpcan, T. & Basar, T., *Stochastic games for security in networks with interdependent nodes*. International Conference on Game Theory for Networks, Istanbul, Turkey, 2009.

[18] Jormakka, J. & Jolsa, J., Modelling information warfare as a game. *Journal of Information Warfare,* **4**(2), pp. 12–25, 2005.

[19] Sallhammar, K. & Knapskog, S., *Using Game Theory in Stochastic Models for Quantifying Security*. Proceedings of the 9th Nordic Workshop on Secure IT-systems, Espoo, Finland, 2004.

[20] MSM, "Making Security Measurable," 8 July 2013. [Online]. Available at: https://makingsecuritymeasurable.mitre.org/.

[21] Alpcan, T. & Basar, T., *A game theoretic analysis of intrusion detection in access control systems.* 43rd IEEE Conference on Decision and Control, Nassau, Bahamas, 2004.

[22] Watters, J., Morrissey, S., Bodeau, D. & Powers, S., *The Risk-to-Mission Assessment Process (RiskMAP): A Sensitivity Analysis and an Extension to Treat Confidentiality Issues.* MITRE, McLean, VA, USA, 2009.

[23] S. Noel, J. S, O. B. & J. M., *Efficient minimum-cost network hardening via exploit dependency graphs.* Proceedings of the 19th Annual Computer Security Applications Conference *(ACSAC'03)*, 2003.

SUSCEPTIBILITY OF IT DEVICES TO HPM PULSE

RAFAŁ PRZESMYCKI & MARIAN WNUK
Faculty of Electronics, Military University of Technology, Warsaw, Poland.

ABSTRACT

The paper presents method for estimate susceptibility of informatic devices for high-power electromagnetic (HPEM) pulses realized by employees of the LAB-KEM MUT. Susceptibility estimation of IT devices will help you understand the susceptibility level of IT devices on the HPEM pulses environment. This estimation focuses on determining the upper level of the IT devices (PC, notebook, HDD, SD card, MMC card, Wi-Fi routers and more) in the HPEM environment.
Keywords: electromagnetic field, electromagnetic pulse, EMC, HPM, HPEM, IT device, susceptibility.

1 INTRODUCTION

The IT infrastructure, personal computers and storage devices are an integral part of our business and everyday life. If you use the IT devices in critical infrastructure or use for the protection, IT security is not acceptable appearance of any disturbances in the electromagnetic environment. In an environment of high-power electromagnetic (HPEM) microwave source is a relatively new kind of risks, both for military and civilian systems. Operation using this type of source is generally referred to as intentional electromagnetic disturbance. HPEM-radiated environment, which is naturally harmful, can also occur at frequencies up to several tens MHz. Generate an HPEM pulse in the direction of the electronic system makes that on the electronic circuits accumulates very high voltage. This phenomenon causes damage to the circuit elements which makes the device unfit for further operation. In the case of a purposeful interaction an impulse is very important in terms of its effectiveness to generate sufficiently strong impulse at the required frequency [1, 2].

Comparatively new risks for military and civil systems in the HPEM environment are high-power microwave (HPM) pulses. An attack with this kind of source is typically defined as intentional electromagnetic interference (EMI). Radiated HPEM environments which are malignant in nature can also occur at frequencies down to a few of MHz.

Technological developments within the area of HPMs have made it possible to produce more powerful microwave sources for illuminating large operational installations. At the same time, it is possible to generate relatively strong fields with compact systems that can be contained, e.g. in a briefcase.

The potential threat from HPM has increased. If an HPM system is fielded by a North Atlantic Treaty Organization (NATO) country, there may be a need to harden co-located friendly systems against the fields produced. However, the actual threat that may exist from friendly or hostile sources in any particular scenario has not yet been fully defined by any NATO nation. The information provided below is therefore general and unclassified. National authorities should be consulted to determine if an HPM environment needs to be specified for any particular system or equipment [3].

While the effects of other HPEMs environment like lightning electromagnetic pulse (LEMP) and nuclear electromagnetic pulse (NEMP) have been tested and measurement in the past and characteristics are readily available, HPM is a comparatively new research area.

DOI: 10.2495/SAFE-V8-N2-223-233

Hostile source characteristics are difficult to predict and possible 'friendly' sources are subject to security issues and have yet to be defined in a meaningful manner, which make defining the HPM environment very difficult [3].

Criminals, terrorists, disgruntled employees/customers, hostile military/special forces and competitors are all potential HPM attackers. The potential scenarios are comparable. A disgruntled employee may bring a small, improvised device into the facility to simply cause some trouble. A terrorist may use a truck-mounted source to cause havoc. A military approach may be to use an airborne system to cause damage [3].

The characteristics of the threat environment are limited by the attacker's technical skill, technology level and access to technological means.

The power of radiated HPM sources ranges from KWs to GWs (peak). In general, the frequency ranges from tens of MHz up to several GHz. Fig. 1 represents a frequency relation between LEMP, NEMP and HPM sources adapted from IEC 61000-2-13 [3].

Depending on the objective and the knowledge of the target, the attacker will select a type of HPM source with the most effective waveform to reach the objective and cause maximum effect. The types of HPM sources and the possible waveforms are described in the standard AECTP-250 Edition 2. The energy levels at the location depend on a wide range of parameters.

Informatic and electronic components and devices (e.g. microprocessor boards) are essential parts of modern civil and military systems like communication, IT-infrastructure or safety systems. Since these informatic components began to control safety critical functions, concern grew over the vulnerability of informatic systems. Therefore the susceptibility of critical systems is of vital interest since a setup or failure in these systems could cause major accidents or economic disasters. The increase of non-metallic materials like carbon-fiber composite as well as the decrease of signal levels results in a decreased susceptibility level of informatic systems. As a consequence, the investigation of the susceptibility of electronic systems as well as their protection and hardening against HPM threats is of great interest.

Due to the design and the functionality of informatic, the employed HPEM test environment and the setup of susceptibility tests, the observed effects differ significantly from each other. In addition manufacturers of informatic systems are reluctant to have the susceptibility data of their systems be published and discussed in public.

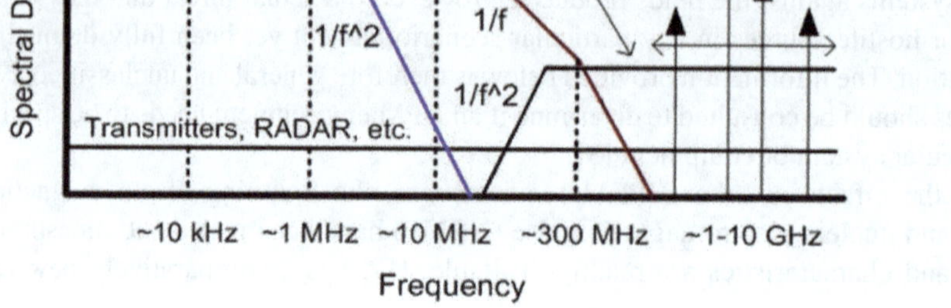

Figure 1: Comparison of Typical High-Power Electromagnetic Environments (HPEM) [3].

2 HIGH POWER ELECTROMAGNETIC PULSE

HPM pulse is a very short duration pulse (the duration time is a few of nanoseconds) and high power (a few of GWs in pulse).

HPM pulses for the first time were observed as a side effect in outbreaks of atomic bombs, it was also found that the extent of such radiation greatly exceeds the means of destruction yet. Currently, several generation methods of HPM pulse were developed through dedicated generators of electromagnetic radiation. Destructive action of E bombs depends on the destruction of all electronic equipment based on semiconductor technology, and in particular cases of destroying whole infrastructure of communications and power transmission. Fig. 2 shows an appearance of HPM pulse in the time domain, while Fig. 3 shows the frequency spectrum of the pulse [4, 5].

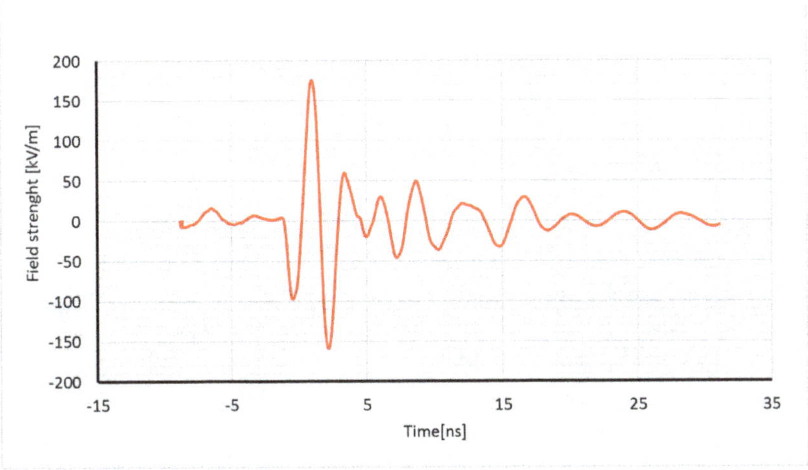

Figure 2: An example of time domain for sample HPM pulse [6].

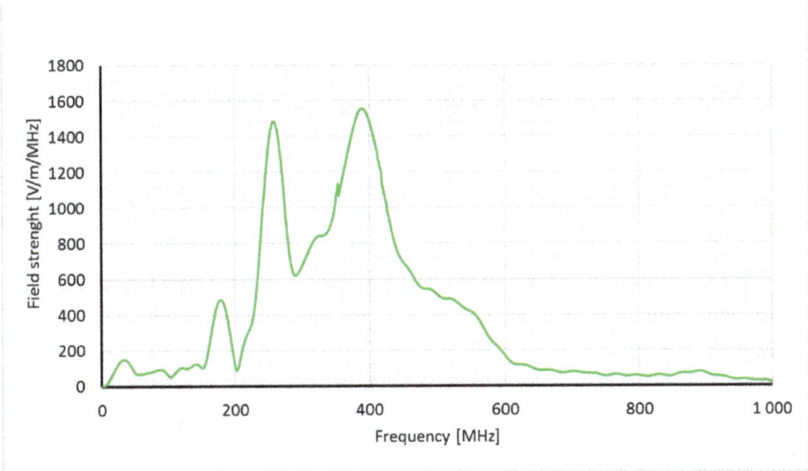

Figure 3: An example of frequency spectrum for sample HPM pulse [6].

3 THE LABORATORY STAND FOR GENERATION AND MEASUREMENT OF HPM PULSE

The main element responsible for the generation of a HPM pulse is a generator Diehl DS 110, which is shown in Fig. 4. It is compact enclosed equipment, looks like a suitcase. Components of the generator are high-voltage power supply, 300 kV Marx generator and a resonant antenna. High-voltage power supply system supplies the Marx generator by 50 kV voltage (symmetrical voltage ± 25 kV), he is charged until the threshold set by the switch that sets pressure gas N_2 in the generator. When changing the position of the switch, antenna is fed a 300 kV voltage, the sparker contained in the air is isolated by the pressure gas N_2 [4, 6]. The basic specification of the HPM DS110F generator is shown in Table 1.

The Diehl Measurement System incorporates a fast digitizing oscilloscope running from an uninterruptible power supply (UPS). The UPS is used in a different manner than originally intended. The UPS operates the oscilloscope for up to 3 hours without mains connection before recharging the batteries of the UPS is required.

Oscilloscope and UPS are installed in a shock damped and electromagnetically shielded housing. This allows not only outdoor measurements without mains available, it also allows

Figure 4: View of the generator Diehl DS 110 [7].

Table 1: DS110F specifications.

Parameter	Value
Size	500 × 410 × 200 mm
Weight	24 kg
Peak radiated power	160 MW
Radiation pattern (without reflector)	Dipole
Pulse duration	4 ns
Repetition rate	>5 Hz (10 Hz typ)
Centre frequency	350 MHz
3 dB bandwidth	100 MHz
Operating time (without reflector)	>1 hours

shielded measurements without any connection to outer environment except the field measuring probe.

The Diehl Measurement System combines a state of the art Tektronix Digital Phosphor Oscilloscope with a computer UPS and an additional battery pack within a shielded and shock damped housing. Due to this combination the Diehl Measurement System can be operated, e.g. in free field without mains, and it can be operated with closed shielding lids when in close distance to radiating HPM systems. The Diehl Measurement System includes a so-called D-dot probe and a balun which are designed to measure pulsed electromagnetic fields. Within the shielded housing a laptop computer is stored usable to calculate measured electromagnetic fields and to further process measured data.

For measuring fast HPM pulses with the Diehl Measurement System a combination of D-dot probe Prodyn AD-70 and a balun Prodyn BIB-170F was chosen as the optimum free field sensor. The probe is mounted on a plastic stand to reduce measurement influences to minimum. Free field sensors as the Prodyn AD-70 are dual sensors. They have two antenna elements that are sensitive to the same field but measure in opposite directions. They have their own ground reference between the two elements. Electric (D-dot) sensors have capacitive sensing elements in opposite to magnetic (B-dot) sensors having inductive sensing elements or loops. The D-dot sensors measure time derivative of electric displacement. The sensor is passive devices and does not need external power supply. Since the sensor is symmetric a balun is required. Balun is the abbreviation for BALanced and UNbalanced and describes a unit which converts the symmetric probe signals to a common, non-symmetric signal as usual in coaxial cables. The D-dot effective antenna size of 0.001 m^2 as well as the balun insertion loss of 8 dB are included as standard parameters within the Excel macros. These values can be changed in Microsoft Excel worksheet when using other probes than original ones delivered with Diehl Measurement System [7].

The main component of Diehl Measurement System is a state of the art Tektronix Digital Phosphor Oscilloscope DPO70404. It contains four analogue channels each having 4 GHz bandwidth and 25 GS/s real time sampling rate. For pulsed field measurement of Diehl Measurement System an interpolated higher sampling rate is not usable, the real time sampling method is always to be used. The oscilloscope is mounted into a shielded 19-inch box using the Tektronix Rackmount Kit. The Tektronix DPO70404 incorporates an own computer board for operating the equipment. It would be possible to install Microsoft Excel with processing macros but Diehl experience is that using a separate computer makes the oscilloscope more stable and data analysis more independent. The oscilloscope allows to record so-called FastFrame formats. FastFrame is able to capture many trigger events respectively many HPM pulses as sequential records in a larger single record. It is possible to view and measure each record individually. Time stamps display the absolute trigger time for a specific frame and the relative time between triggers of two specified frames. This means that several HPM pulses can be stored within one file and can be processed in the computer in one step. Data measured with the oscilloscope can be stored either on internal hard disk, on USB sticks connected at back or front USB ports or directly on network drives when connected with other computers. Diehl recommends to store measured data first on internal hard disk. To copy them onto the laptop computer use either USB memory sticks or network connection. This insures that only a duplicate copy is onto the laptop computer when processing measured data [7].

The output of the D-dot sensor is proportional to the first-time derivative of the electric displacement D(t), that is:

$$\text{D-dot} = dD/dt \tag{1}$$

and, rewriting equation:

$$Vo = R\ Aeq\ dD/dt, \text{ or } dD/dt = Vo/R\ Aeq \tag{2}$$

then, recalling that $D = \varepsilon o\ E$,

$$dE/dt = Vo/R\ Aeq\ \varepsilon o \tag{3}$$

The output of the D-dot sensor is measured in volts that are proportional to units the first time derivative of electric displacement dD/dt, coulombs per square meter. The constant of proportionality between the sensor output and the first-time derivative of the electric field dE/dt (= dD/dt εo) is the permittivity constant εo. The constant of proportionality between the sensor output Vo and the first-time derivative of the electric field dE/dt is the product of the output impedance R and the equivalent area Aeq of the sensor, and the permittivity constant εo. This is the calibration factor, but it is applied to the first time derivative of the electric field, dE/dt, not to the electric field itself [8].

The sensor output resulting is a cosine function with amplitude Vop and frequency f, where:

$$Ep = 1/2\ f\ (dE/dt)p = Vop/2\ f\ R\ Aeq\ \varepsilon o \tag{4}$$

In the sinusoidal or CW case, the peak value of the input field can be obtained from the peak value of the sensor output, the constant of being proportionality being the product of the general constant R*Aeq o and 2*f, which is constant if the frequency is constant when the measurement is made, as in the CW case.

A reasonable estimate of the maximum value of D/t or E/t is obtained by measuring the change in D or E over the time interval during which D or E is changing most rapidly. This is usually taken as the T/6 duration when D or E is going from +0.5 (Dp or Ep) to -0.5 (Dp or Ep). This is the maximum output we expect from the sensor.

The integrated sensor output is obtained by integrating the sensor output with respect to time. If the sensor output has been digitized, the integral can be obtained easily by multiplying the numerical values of D/t by t over the time interval 0 to t. As the number of discrete points increases, D/t becomes a better estimate of dD/dt and (D/t) t becomes a better estimate of dD = D = 1/Aeq* Vodt [8].

Fig. 4 has been presented a block diagram of the laboratory stand to generation and measurement of HPM pulse. The main elements of this laboratory stand are the Diehl generator and field probe, which they are placed in an anechoic chamber so that the probability of damage other electronic devices located near the generator is very small. In addition, they have been placed on the metal floor with ferrite absorbers thus ensuring that a common reference mass. Devices in the measurement room allow their safe operation by authorized persons [1, 9].

Also in this case a common reference mass of the measuring devices is ensured by a protective metal casing. Components of the laboratory stand located in the anechoic chamber arranged in accordance with block diagram of Fig. 5 are shown in Fig. 6.

4 HPM SUSCEPTIBILITY TESTING

Estimating the susceptibility levels for the IT equipments is to determine the level of susceptibility to the impact of this device by pulsed electromagnetic field. Estimation of susceptibility levels will be presented on the example of the few chosen IT equipments. We are selected

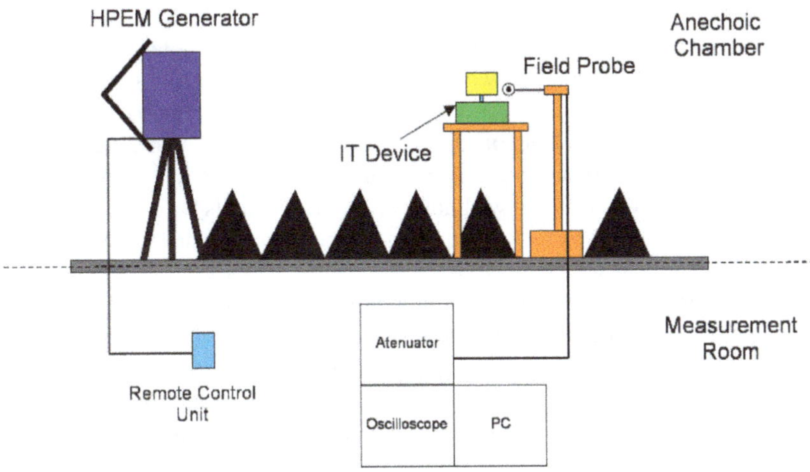

Figure 5: The block diagram of laboratory stand for generating and measurement of HPM pulse.

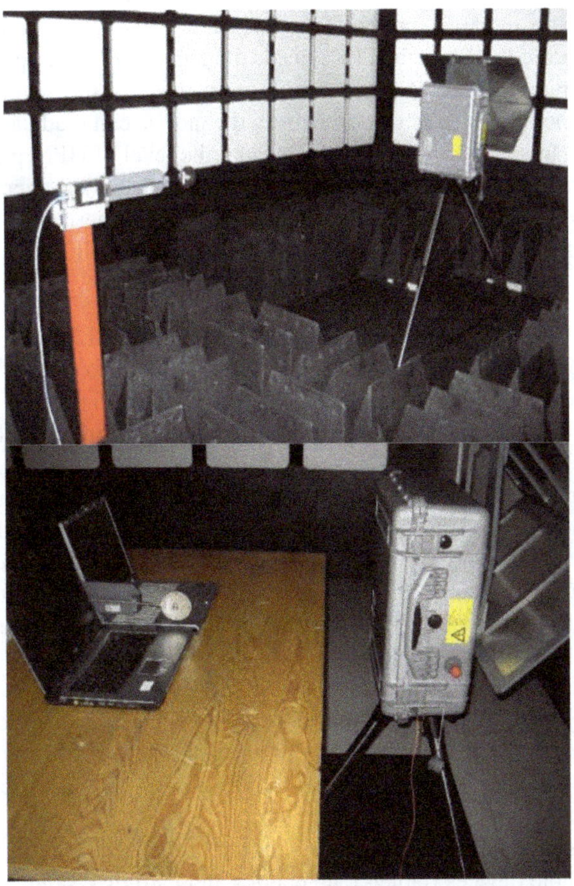

Figure 6: The view of laboratory stand for measurement of HPM pulse.

notebook computers, tablets, TEMPEST computer and Wi-Fi routers external hard disk, Flash memory and memory cards. For testing was prepared eight notebook computers, four tablets, two TEMPEST computers, two Wi-Fi routers, four hard disks, six memory cards (four SD cards and two MMC cards) and five USB Disk (Flash memory).

Before proceeding to set the level of susceptibility, we must apply to the characteristics of the selected device. The more data we collect about the devices we will more precisely determine the level for fulfillment. When we know the all the elements we can affect the levels of susceptibility/immunity of informatic equipment to the impact pulsed electromagnetic field, we can proceed with the research [1, 4].

The measuring process is start from selection of informatics device (PC, tablet, TEMPEST PC) and run or start from writing to all data storage media files with information containing diverse content. In this experiment, data storage media stored text file created in MS Word, the file containing the image in jpeg format, binary file with the extension bin and the file with the recorded film with extension avi. After saving data on individual data storage have been reading validation of individual files on all data storage equipment [1].

Then, the chosen IT device is placed into anechoic chamber, which there is a laboratory stand for generation and measuring of HPM pulse. After placing the IT device into an anechoic chamber we can starts the triggers HPM pulse generator. In the measurements we set to triggering the generator 1 HPM pulses in one second, and 10 HPM pulse in one second. After the process exposing HPM pulses IT device, the device is checked with the other computer is working correctly. When the IT device is damaged and it is impossible to turn on it, we determined immunity level for the IT device on the basis of the generated field strength. In the case where the IT device is working correctly and we can read data stored on device, we put it back into an anechoic chamber, increasing the level of HPM pulse and we release the generator, thereby exposing our IT device by HPM pulses with increased power. After exposure again we check the correct operation of our device. These steps we repeat until we set the susceptibility levels on pulsed electromagnetic field for all tested IT device [1, 4].

5 OBSERVATIONS AND TEST RESULTS

The study consisted in verifying the behavior of the device information in the high-power pulse electromagnetic field. The investigated object (information device) was placed in an anechoic chamber 80 cm above the ground and moved away from the HPM generator at different distances, resulting in different field strength [1].

In the experiment the research were subjected to chosen IT devices that shortly characterized in Table 2. The chosen devices were exposed to 1 HPM pulse and 10 HPM pulses. The rest of this article presents only the results of the exposure of IT devices for a 1 pulse, due to the fact that exposures computer equipment for a 10 HPM pulses, the results were identical. It can be concluded that no number of pulses but amplitude of pulse in does case the interference (damage) of the device [1].

Table 3 summarizes the susceptibility of different IT devices on HPEM fields. This table shows the statistics on the number of interference (damage) of the chosen IT devices in relation to the level of intensity HPM pulse. For example, the value of 3/5 indicates that the test were three interference (damage) on a five series of exposures consisting of HPM pulse [10, 11].

The statistics concerning a susceptibility of chosen IT device show that most failures were Wi-Fi routers, portable devices (tablets), notebooks and hard disk's, and the least damage is

Table 2. Characteristics of selected IT devices being tested.

IT device	Description
PC 1 to PC 8	Notebook computer manufactured in 2015 and 2016 year
Tablet 1 to Tablet 4	Tablets manufactured in 2015 and 2016 year
TEMPEST	Two TEMPEST Computers manufactured in 2015 and 2016 year
Router 1 and 2	Router WLAN 54MBits manufactured in 2015 and 2016 year
Hard Disk HDD 1	External Hard Disk Drive 20 GB, USB 2.0 interface with an additional power cable
Hard Disk HDD 2	External Hard Disk Drive 40 GB, USB 2.0
Hard Disk HDD 3	External Hard Disk Drive 100 GB, USB 2.0
Hard Disk HDD 4	External Hard Disk Drive 1 TB, USB 3.0
Memory Card SD 1	The SD memory card with a capacity of 16 MB
Memory Card SD 2	The SD memory card with a capacity of 16 MB
Memory Card SD 3	The SD memory card with a capacity of 512 MB
Memory Card SD 4	The SD memory card with a capacity of 1 GB
Memory Card MMC 1	The MMC memory card with a capacity of 32 MB
Memory Card MMC 2	The MMC memory card with a capacity of 1 GB
USB Disk 1	External memory type Flash Drive 128 MB
USB Disk 2	External memory type Flash Drive 256 MB
USB Disk 3	External memory type Flash Drive 512 MB
USB Disk 4	External memory type Flash Drive 1 GB
USB Disk 5	External memory type Flash Drive 2 GB

Table 3. Test results statistics for selected data storage devices.

	10 kV	30 kV/m	60 kV/m	80 kV/m	120 kV/m	180 kV/m	250 kV/m	320 kV/m
PC 1	0/5	3/5	4/5	5/5	5/5	-	-	-
PC 2	0/5	3/5	4/5	5/5	5/5	-	-	-
PC 3	0/5	3/5	3/5	5/5	5/5	5/5	-	-
PC 4	0/5	2/5	4/5	5/5	5/5	5/5	-	-
PC 5	0/5	2/5	3/5	5/5	5/5	-	-	-
PC 6	0/5	2/5	3/5	5/5	5/5	5/5	-	-
PC 7	0/5	3/5	4/5	5/5	5/5	-	-	-
PC 8	0/5	3/5	4/5	5/5	5/5	5/5	-	-
Tablet 1	0/5	3/5	3/5	5/5	5/5	-	-	-
Tablet 2	2/5	4/5	4/5	5/5	5/5	-	-	-
Tablet 3	1/5	3/5	4/5	5/5	5/5	-	-	-
Tablet 4	2/5	4/5	4/5	5/5	5/5	-	-	-

(Continued)

Table 3. (*Continued*)

	10 kV	30 kV/m	60 kV/m	80 kV/m	120 kV/m	180 kV/m	250 kV/m	320 kV/m
TEMPEST 1	0/5	0/5	0/5	0/5	1/5	1/5	4/5	5/5
TEMPEST 2	0/5	0/5	0/5	0/5	1/5	1/5	3/5	5/5
Router 1	4/5	-	-	-	-	-	-	-
Router 2	5/5	-	-	-	-	-	-	-
HDD 1	0/5	0/5	4/5	5/5	5/5	-	-	-
HDD 2	0/5	0/5	0/5	0/5	4/5	5/5	5/5	-
HDD 3	0/5	0/5	0/5	0/5	4/5	5/5	5/5	-
HDD 4	0/5	0/5	0/5	0/5	0/5	5/5	5/5	5/5
SD 1	0/5	0/5	0/5	0/5	0/5	0/5	0/5	0/5
SD 2	0/5	0/5	0/5	0/5	0/5	0/5	0/5	0/5
SD 3	0/5	0/5	0/5	0/5	0/5	0/5	0/5	0/5
SD 4	0/5	0/5	0/5	0/5	0/5	0/5	0/5	0/5
MMC 1	0/5	0/5	0/5	0/5	0/5	0/5	0/5	5/5
MMC 2	0/5	0/5	0/5	0/5	0/5	0/5	0/5	5/5
USB Flash 1	0/5	0/5	0/5	0/5	0/5	1/5	1/5	1/5
USB Flash 2	0/5	0/5	0/5	0/5	0/5	1/5	1/5	1/5
USB Flash 3	0/5	0/5	0/5	0/5	0/5	1/5	1/5	5/5
USB Flash 1	0/5	0/5	0/5	0/5	0/5	1/5	1/5	5/5
USB Flash 4	0/5	0/5	0/5	0/5	0/5	1/5	1/5	5/5
USB Flash 5	0/5	0/5	0/5	0/5	0/5	1/5	1/5	5/5

the SD memory card and TEMPEST Computer. Immunity levels for each IT devices are as follows:

- for SD and MMC memory cards – above 320 kV/m,
- for TEMPEST computers – 320 kV/m,
- for flash memory (USB flash drive) – 250 kV/m,
- for notebook computers – 180 kV/m,
- for hard disk's – 170 kV/m,
- for tablets – 120 kV/m,
- for Wi-Fi routers – 30 kV/m.

The results of the statistics also show that the newer media (higher capacity, higher cycle frequency the processor clock) were much less susceptible to interference than the older hardware HPM.

6 CONCLUSIONS

The possibility of an attack on information technology equipment which work only on a specific frequency significantly restricts the use of this equipment in critical infrastructure. Generate an electromagnetic pulse located in the working range of informatic devices can result burnout internal circuits which will make it will be completely unsuitable for further work. Remember to adequately protect the informatic device [1, 4].

The article presents the susceptibility levels of selected IT devices on exposure of HPM pulses and set immunity levels on HPM pulses of selected IT devices at which the devices have been destroyed.

Please be aware that the results presented in the article are very limited and we should collect a more data before the data presented will be introduced into everyday life. They are also required further research to determine, in more detail, of susceptibility data storage equipment based on a sample consisting of a larger number of devices available on the market.

REFERENCES

[1] Przesmycki, R., Wnuk, M. & Bugaj M. – *HPEM susceptibility assessment of data storage devices,* 2016 21st International Conference on Microwave, Radar and Wireless Communications (MIKON), 9–11 May 2016, Kraków, Poland, Electronic ISBN: 978-1-5090-2214-4.
https://doi.org/10.1109/MIKON.2016.7492113

[2] Kubacki, R. & Wnuk, M. – *Broń elektromagnetyczna – broń przyszłego pola walki, Nowoczesne technologie systemów uzbrojenia,* Red. Z. Mierczyk, WAT, Warszawa , pp. 461–470, 2008.

[3] AECTP 250 Ed.2 2011 : *Electrical and electromagnetic environmental conditions,* NATO INTERNATIONAL STAFF - DEFENCE INVESTMENT DIVISION, 2011.

[4] Przesmycki, R. & Wnuk, M. – *HPEM susceptibility assessment of informatic devices,* Progress in Electromagnetic Research Symposium (PIERS), 8–11 August 2016, Shanghai, China, 978-1-5090-6093-1.
https://doi.org/10.1109/PIERS.2016.7734568.

[5] Przesmycki, R., Wnuk, M., Nowosielski, L., Piwowarczyk, K. & Bugaj M., *The Shielding Effectiveness Measurement Using High Voltage Pulse Generator,* Moscow, Russia, PIERS Proceedings 2012, str. 1449–1452, ISSN 1559–9450.

[6] Przesmycki, R., Nowosielski, L. & Wnuk, M., *Metodologia pomiaru absorpcji materiałów z wykorzystaniem generatora HPEM,* Przegląd Telekomunikacyjny, ISSN 1230–3496, NR 8-9/2014, str: 821–827.

[7] DIEHL BGT Defence – *Operating and Maintenance Manual for DIEHL DS110. KS2.F.MP1.B (HPM Case System DS110),* document No BDB8118952.

[8] Reed Edgel, W., *The next Best Thing to Calibration, Functional Testing,* Prodyn Aplication note.

[9] Rafał, P., *Możliwości zastosowania energii skierowanej do niszczenia urządzeń informatycznych,* KSTiT Kraków, Polska 16 - 18.09.2015r., Przegląd Telekomunikacyjny, ISSN 1230-3496, e-ISSN 2449-7487, NR 8-9/2015, str: 1178–1181.

[10] Hoad, R., Lambourne, A. & Wraight, A., *HPEM AND HEMP susceptibility assessments of computer equipment,* 17th International Zurich Symposium on Electromagnetic Compatibility, pp. 168–171, 2006.

[11] Brauner, F., Sabath, F. & Haseborg, J.L., *Susceptibility of IT network systems to interferences by HPEM,* 978-1-4244-4267-6/09/$25.00 ©2009 IEEE, pages 237–242.

SECURITY ASSESSMENT CASE STUDIES OF PUBLIC BUILDINGS IN INDIA

MANJARI KHANNA KAPOOR
School of Planning & Architecture, New Delhi, India.

ABSTRACT

Security engineering is a valuable tool in the design of our built environments and almost a compulsory inclusion in some parts of the world. At the same time there is a wide range of approaches and techniques with their own efficacies and vulnerabilities. A lot of the so-called security around us is mere theatrics and many of the urban environment design decisions stem from perceptions and heuristics. A careful scientific analysis and systematic assessment of 'secured environments' helps identify the true merit of many such tools. The aim is to reduce the subjectivity in security and objectify it. Some such experiments with audit of security led us to some startling revelations indicating a direct relationship between security and common sense. We have evidence to deal with the myths and reaffirm the direct relationship between good architectural design and sensible security. The effort is to snatch our built environment from the clutches of the security men and design-in a more wholesome version of security. The idea is to accept and include security as a necessary component, as essential as any other service and integrate it into our built environment seamlessly.

KEYWORDS: security, assessment, engineering, crime-prevention, counter-terrorism, designed-in, resilience, organised, mechanical, natural, costs.

1 INTRODUCTION

Security is both a physical necessity and a psychological need today. We see a lot happening around us in the name of security from mechanical militarization to placebo theatrics. There is a lot that is technically feasible with growing technological inventions and design innovations. There is lot of literature on these and many guidelines formulated across the world to guide the designers of the urban world.

Perception of the security however is a physiological and psychological necessity but subjective to individual thresholds and cultural predispositions. A doctoral survey by the author lays bare the socio-economic and cultural factors that influence the security perceptions. It establishes a certain link between security perceptions and the cultural predispositions, theological standpoint with a dash of socio-economic markers. This is one of the biggest challenges for the security engineers and potential designers of a secure environment.

As designers and engineers we need measurable objectives, quantifiable efficacy to target and eliminate the eyewash. It is also valuable for justification of project expenses and value based engineering. With the objective of ridding security of its subjectivities, the author's doctoral research undertakes case studies of some highly secured environments (in India). This led to some interesting revelations and greater understanding of the security engineering approaches.

They help us go beyond the perceptible top layer into the deeper realms of well-founded security approach. These case studies reveal a number of factors at play which contribute to security or take away from it intentionally or accidentally. Some of these have larger ramifications for the architectural design paradigm and ideology in general.

DOI: 10.2495/SAFE-V8-N2-234-245

2 METHODOLOGY

This paper undertakes close scrutiny of two public buildings in India to study, understand and analyse in depth from the security standpoint. Both the samples are very successful in their security provisions and selected on the basis of:

- high population count or heavy footfalls;
- high threat profiles;
- high security inputs in design and operations;
- high security perceptions attached to the need & implementation

Based on the above rationale, we identified 2 buildings exemplifying completely different building function and type of security designed. The first is a temple and the second an office complex each with their unique complexities. Their specific details are discussed in the next section.

The systematic methodology for assessment started with some cursory knowledge about these installations, backed by a preliminary literature survey. This was then substantiated with a fair amount of fieldwork including repeated visits to the site, observations and interviews. Simultaneous analysis of the drawings and details entailed a lot of back & forth. All the assimilated queries were then discussed with the security in charge, policy makers, designers and the project team to get a complete representation of the intent and operations.

All the observations were then carefully analysed for the related factorials. The assessment takes into account the various aspects of security including counter-terrorism, overall crime prevention through environmental design (CPTED) and other related factors like resilience.

The assessment deconstructs the overall security into smaller identifiable components such as perimeter protection, access control, surveillance, vehicular control, MEP design, structural inputs, disaster resilience, design strategies and almost every related aspect as researched. These are then further segregated under the heads of Natural means through design, resilience, inputs aimed at crime prevention and lastly those which are merely serving to feed the perception component without actual effect. This analysis gives an overall picture of the soundness of the security efforts, their appropriateness to the building typology, their cost justifications and efficacy in achieving their security goals.

Please note that Natural security refers to the components which secure by the effective use of design be it spaces, materials or surveillance without the intervention with mechanical (equipment based) or organised (personnel or guards) measures.

So each of the identified buildings was first carefully examined in their layouts and literature survey to get an understanding of the history and context. This, supplemented with interviews with the security designers of the premises help identify the exact list of threats to deal with, which is essential to identify the best strategy in design. Usually this is the best stage to undertake a careful field observation of the working of the building. All questions that arise from the exercise are then addressed by the many interviews with the project team including the architects, structural engineers, MEP designers and the policy makers.

All the collected information is then uploaded and broken down into the 4 categories as listed above to throw up a clear proportion of each aspect at play in securing the premises. Also important is the effectiveness and performance of each of these, derived and analysed in the psycho-social contexts too.

3 CASE STUDIES

3.1 THE SWAMINARAYAN AKSHARDHAM TEMPLE, New Delhi, India.

The Akshardham temple in New Delhi is not as much a Hindu temple as it is a spiritual-cultural campus. More than its religious relevance, the temple is a centre for tourists to gain awareness of the Indian culture and tradition attracting more than '70% of the tourist population to the capital city' (www.akshardham.com [1]).

3.1.1 Security context

The 'Bochasanwasi Shri Akshar Purushottam Swaminarayan Sanstha' (BAPS) owns and patronizes many such temples across India and overseas. While this location was still under construction their campus at Gandhinagar, Gujarat in India was attacked in September 2002 suffering casualties of approx. 30 pilgrims and few security men (Patel [2]).

Traditionally, India is a country with a long history of terrorist attacks and anti-terror security regimens. Also the Hindu–Muslim friction is nothing new especially in the backdrop of the communal tensions that have remained antipathetic ever since the partitioning of the country on theistic basis. Even today, theological wars are raging in the valleys of Kashmir which remains a territorial and a contentious issue for India and its breakaway neighbours accused of harbouring terror for political interests.

In that context, a Hindu temple carries an inherent vulnerability in the current times. Add to that the tourist interest and population numbers. All these continue to put this temple as one of the prime targets in the capital city state. This has made the establishment most aggressively conscious of its security design, sometimes even at the cost of many a tourist getting intimidated by the coldness of the security checks in a place that one would seek calm and inner peace.

3.1.2 Threats Identified:

AS PER THE Public Relations Officer (PRO) for the premises, Shri Harish Patel, the chief threat perceived by the administration is 'the unpredictability of the human being' (Patel [2]). This includes human volatility and even small feuds and common crime of any sort that consecrates the environment and spiritual ambience.

AS PER THE AUDIT: Terror, blast, petty crimes, armed attack or hostage situations, bio-chemical attacks, civil unrest, and sabotage in addition to the usual risk of fire and other natural disasters. Being a public place with a huge site setting, gardens etc., women's, kids or seniors' security can be an additional concern depending upon the architectural design.

3.1.3 Occupants:

The visitors are the main thrust of this architectural accomplishment including the faithful pilgrim, the curious Hindu outside the immediate cult, the non-Hindu tourist and the International seeker of Indian culture and tradition. The numbers on a given day can go up to a lakh or beyond on national holidays. The staff of the Trust, volunteers, administration and service staff also constitute a sizable population including facilities for their smooth operations.

3.1.4 Design overview:

'The temple, at the centre of the complex, was built according to the *Vastushastra* and *Pancharatra Shasta*.The complex features an Abhisheka Mandap, Sahaj Anand water show, a thematic garden and three exhibitions namely Sahajanand Darshan (Hall of Values), Neelkanth Darshan (an IMAX film on the early life of Swaminarayan as the teenage yogi, Neelkanth), and Sanskruti Darshan (cultural boat ride). According to Swaminarayan Hinduism, the word *Akshardham* means the abode of God and believed by followers as a temporal home of God on earth.' (www.akshardham.com [1])

Designed by an in-house team of volunteers (swamis), a local architect was appointed for the liaison works only. The security also is designed mostly in-house under the guidance of the Head of Security, a retired Brigadier from the NSG.

This temple has been recognised by the Guinness World records as the *World's Largest Comprehensive Hindu Temple*. The certificate states, 'BAPS Swaminarayan Akshardham in New Delhi, India, is the world's largest comprehensive Hindu temple. It measures 356 ft. (109 m). Long, 316 ft. (96 m). wide and 141 ft. (43 m). high, covering an area of 86,342 sq. ft. (8,021.4 m^2). The grand, ancient-style, ornately hand-carved stone temple has been built without structural steel within five years by 11,000 artisans and volunteers.' (www.akshardham.com 2017) The complex also houses a guest-house for visiting members, volunteers and swamis. In all there is accommodation for 850 volunteers in approx. 224 family dwelling units' incl. 108 kids (Patel [2]).

The main core is a predominantly single storey structure built as per the Hindu tradition following age-old concepts of site planning which have served well for security and urban design since generations. However, the ancillary functions follow the traditional look with modern technology.

3.1.5 Site planning:

The choice and location of site is its greatest element, being quite distant from the main city centre, flanked on two sides by arterial roads maintaining more than 100 metres distance on at least two sides from any of its neighbours. The neighbours too are of a nature that do not contribute to the vulnerabilities and are in turn least affected by the temple itself.

On the balance two sides the temple is served by a Metro station and a bus terminal though well distanced and yet integrated in the urban design. On the 4th side (towards the river banks) lies a residential township which could have been a vulnerability but it is well anticipated in design though it came up a few years after the completion of the temple.

The functions are strategically zoned in layers. This strategy is reminiscent of the traditions from historical monuments like Indian forts and palaces (see Fig. 1 above).

- PERIPHERAL Public zone – public and less secured periphery used for roads and other non-secure peripherals adding to the stand-offs distances from roads.
- SEMI-SECURE Private zone – accessible by authorisation only for Intermediate functions & backend support.
- SECURE Public zone – this is the main core including the shrine and the human surge at the heart, accessed only after due sanitization and under strict scrutiny and surveillance.

This arrangement of public operations is completely introverted exposing only a limited part of the property to casual scrutiny. Services buildings and staff accommodations are adequately buffered from all public areas with strictly regulated internal traffic circulation. These

Figure 1: Satellite image of the Akshardham temple (maps [3]).

also add to the protective layer distancing from the side with unknown future development. Parking is strategically distanced from the secured core, on ground but detached.

3.1.6 Perimeter:

"Security is not importantly by boxed in walls but mobility of public with assurance of safety. So comfort included in security"- (Patel [2])

Boundary wall (10ft or more) further mounted with a concertina coil (Fig. 2 below) convey the look of an embattlement more than a tourist destination. A more inclusive and transparent perimeter may have been more conducive to the use category but since it is not a part of a larger urban design, it works despite the hostility it generates. The boundary wall supplements the ravine trench on two sides making the severity of the wall seem overdone.

The perimeter is over hardened while it was an opportunity to create a friendlier and magnanimous look and integrate it with the urban design including the Metro station and bus terminus. The hardening does create a sense of suspense and mystique but also an intimidating dent in the peaceful ambience desired and that lies beyond this point.

Figure 2 and 3: Boundary wall and gates to the temple complex.

3.1.7 Access control:

Large and sturdy gates are the main measures to control access. They are rendered in the historic and traditional feel (see Fig. 3 above). There are six gates to the complex:

- Gate 1 – open to public vehicles
- Gate 2 – VIP gate only operated for VIP movement
- Gate 3 – service entry for the cafeteria & guesthouse kitchens
- Gate 4 – entry to the residential part- not accessible to public vehicles
- Gates 5 and 6 – entries from residential block to the temple complex.

These gates are each guarded by personnel 24 x 7 and further hardened by boom barriers or tyre busters to protect against forced entry. It may be noted that a large part of these measures are mandatorily stipulated for public places in India. However, in this temple we find them well integrated in design without inconveniencing the visitors or the neighbours.

3.1.8 Equipment:

A number of temples and other tourist places in India do not permit carrying phones and other personal belonging due to constraints on checking and security protocols. These restrictions cause all kinds of inconveniences to people creating a parallel terror profile of its own where every individual needs to be forewarned and ready to surrender many of their belongings.

The temple being studied is one of the few that complements its stringent restrictions with amply provided cloakrooms and lockers, free of cost to all and sundry irrespective. *'Security by providing facility- 'decently humanly feasible'* (Patel [2]) is part of the BAPS ideology.

These cloakrooms and structures within which the security scanning and frisking takes place, prior to the main entry and temple complex are detached from the rest of the grandeur and built of temporary construction. This makes them seem like an afterthought but actually they fare better due to their low damage potential.

After providing adequate facilities to the visitors the temple administration is very thorough with its scanning, screening and if required frisking to almost the levels followed at international airports. This is a done thing for the Indian tourist and almost a pre-requisite to most locals to feel secure in an environment. However, the research witnessed many tourists getting spurned by the severity at this point. They consider it as a glimpse of further militarization inside and many times opt out, little realising that once the visitor is sanitised the environment inside is one of spirituality and peace.

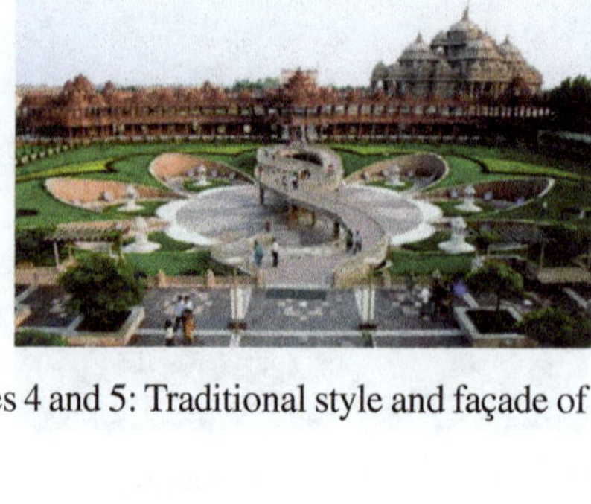

Figures 4 and 5: Traditional style and façade of the temple complex (www.akshardham.com [1]).

3.1.9 Surveillance:
The peripheral areas are open and offer unhindered surveillance by natural and mechanical means. The semi-secure however being triple storeyed needs more mechanical means but the secure zone is laid out in the traditional style offering grand opulent plazas and landscapes which facilitate natural surveillance 24 x 7 (see Fig. 4 below). This is further substantiated with a generous distribution of volunteers across the entire complex.

This appears to be an out-dated strategy but the manner and subtlety of its application here actually makes it very appropriate. The volunteers are mostly theistic devotees of the cult, educated and well trained individuals who are not armed security men. So these people while offering casual assistance to the tourist also are the eyes on the street and the secondary function at work discreetly keeping an eye on every activity, foretelling an ugly situation, exercising crowd management, cleanliness and maintenance, all at once.

3.1.10 Resilience:
The design is mostly traditional for the Hindu ambience and this too performs well from security view point (see Fig. 5 above). It allows low rise traditional development without structural risks constructed with the historical technology, which has been holding up many old temples for the last many centuries now, even in the face of seismic and other natural forces.

These stone structures have less vulnerability and their external facades are less susceptible to splintering or flying debris. Further the Hindu traditional system follows less openings with the few also screened with jaalis in response to the climate needs. This vocabulary fares extremely well security wise too, offering little insight into the interior functions for better privacy.

The complex is designed in a manner that requires least dependence on the MEP, structure and other disaster strategies of evacuation and egress. This owes largely to the traditional type of layout and construction which further integrate the disaster resilience, fire and life safety codes rather seamlessly. Most other contemporary good practices are well integrated into the operation module which make this tourist destination notch up further on its security audit.

3.1.11 Main findings of the case study:
The Akshardham temple is a non-critical public building falling in the medium threat category as per the building type but is treated in design as a medium to high risk category due to high risk perceptions.

The security thought is of a high degree which seems too harsh at places in the context of an otherwise peaceful and calm environment. The complex typifies the psychological play through security design and its appropriate use especially in public places serving cross cultural targets. It needs to accept and understand that globally there can be many interpretations to the same security. At the same time it also exemplifies the universality of the natural means of security as the most effective, non-controversial and non-subjective form.

The traditional approach is a very successful technique to achieve the natural security in planning, finishes, structural and resilient aspects. Even new age concepts of CPTED are well steeped in the traditional process. Essentially the traditional system hinges on simple common sense response to a problem and is hence timeless as it is cost effective. Simple strategies of functional layering, distancing vehicles, seamless screening of huge numbers, their guided management and other covert strategies are found to be much more efficient than the mechanization prevalent in our urban world today.

3.2 DATA CENTRE COMPLEX for the government of India at Bengaluru, India.

The Government of India undertook the establishment of its Data centres to house the data of the projected 150 million population. Since they would house sensitive data and being a project of national prestige the complexes were perceived as being 'critical', highly sensitive and with a high threat profile. As a result they came to symbolise a Nation's IT prowess and Engineering muscle in the face of the antinationalistic malicious agendas. (Aggarwal [4])

The campus is designed to house a staff of approximately 1000–1500 within, of these about half would be working 24x7. Being a public utility, it is likely to attract a visitor population of thousands daily. The small museum within is likely to attract busloads of students and tourist excursions. (Grover [5])

3.2.1 Site planning

The site selection was not easy in the heart of the burgeoning IT city. Though it is an abundant site, its proximity to a residential colony and access through it creates insurmountable vulnerabilities. Plotted houses touch the site boundary on 1 complete face while 2 faces are vacant for future development. There future use will define another level of threats in the future that cannot be identified now.

The campus is divided into 3 building blocks in order of security concerns (refer Fig. 6 below)-

- Data centre block housing the data storage servers considered the most critical assets.
- Non- data centre block houses the support staff, administrative and public utilities. These are non-critical functions but still high security areas due to the human collateral.
- Utilities block accommodates all services required to make the installation function non-stop. This also has an element of criticality as a breakdown of this support will hinder the critical functioning of the Asset. (Grover [5])

Setbacks and peripheral fire tender driveway is the only vehicular movement The Data centre is tucked back with restricted vehicular movement around it. The strategy to recess the DC block to the rear to layer it and create more distance is exposing it to the vulnerability of the adjoining residential plotted housing. Hence it is designed to neutralise the locational vulnerabilities. (Grover [5]). This block has no basement and is structurally detached. It is

Figure 6: Data centre complex at Bengaluru – layout plan

rendered as a closed cuboid with the outer skin wall designed as a 600mm thick RCC wall, independent of the internal structure with an additional column and slab frame

The NDC block sits on a single basement parking and some services to keep it functional and independent of the critical services block. It houses 6 floors of Tech centre, Regional office, Guest house and accommodation for security men.

3.2.2 Perimeter, access control and equipment:

The boundary wall on 2 sides is 3m high with a concertina coil. On the front it gets relief with solid steel gates and some relief design and finishing. The 4th side adjoining the neighbours however has a 12m high fence atop the boundary wall (see Fig. 8 below). All perimeter openings are further hardened with boom barriers and guard rooms with latest scanners and authorisation facilities.

The peripheral plantation type, height, diameter of foliage is all controlled as per the security inputs. Further some planters and mounds are so created in line with Security strategy to resist forced entries. (Aggarwal [4])

The site has 3 main gates. 2 on the main access face including 1 each for entry and exit. The third one located on the residential face is dedicated to service movement and authorised staff entry only. This gate is provisioned with parking and screening of service vehicles and their cargo. The authorised staffs however are allowed to zip through to reach the basement or the porch where they are intensively screened. (Aggarwal [4])

The visitors use the front gates which are designed to facilitate queuing and space for checking of each and every vehicle, its contents and passengers. The vehicle and its contents are sanitised before admissibility into premises. (Grover [5]) The strategy is meant to neutralise the vulnerability of getting the vehicle into the structure be it the porch or the basement. The spatial constraint necessitates the mechanization of security and maybe justifiable in a critical installation of this nature.

Though most of the invasive checking is done at the gate itself, baggage and personnel screening is mandated at the porch drop-off and the lift entrances in the basements. The vertical circulation within has too much multiplicity of entry and exits thereby exaggerating the security inventory. A lot of the circulation has been curtailed from security considerations.

3.2.3 Surveillance

The surveillance strategy is based mostly on cameras of the latest technology and monitoring in the 24x7 security station located very focally at the lobby level. This unit is operational round the clock and supported by staff dormitories, change rooms, lockers and toilets on the top floor of the NDC block.

Though the NDC block has a glass façade (see Fig. 7 below) and the office spaces inside are designed to look out there is not much spatial scope for surveillance because of the architectural form in the site shape. The DC block is as it is opaque (see rear part in Fig. 7) but the residential neighbours offer an insurmountable vulnerability by offering well concealed vantage points from the outside. The only good part is that here is not much that they can damage directly since the DC is so solid.

3.2.4 Resilience:

Resilience is probably the greatest strength of the DC block with high levels of thought to structural, MEP, materials, fire and life safety. The NDC block on the other hand is restricted

Figure 7 & 8: Data centre complex at Bengaluru – façade and fencing towards residential side.

to only codal compliance to structure, disaster and life safety. The glass is not ideal but still safe enough due to its compliance to green standards. The lamination makes it less dangerous. The granite stone on the façade and other façade finishing materials are regionally sourced but have a high potential to harm.

The DC block however is designed to ensure that even in the worst case scenario the critical operation can continue unabated. The MEP design includes redundancy of all services critical to the data centre including power back-up, air conditioning, lighting *et al*. The structural hardening includes a 600 thick RCC wall designed to sustain blasts up to 5 kg of TNT at a distance of 20m and detached from the framing of the floor slabs within. The structural design and construction has ensured all joints and structural members are built for the blast shear including even the roof (Singh, 2016) The façade does not have more than 1sqm of openings in keeping with the blast resistant requirements. All doors are shielded by RCC aprons or vestibules. (Singh [6])

3.2.5 Main findings of the case study:

The Data centre Complex at Bengaluru is a high risk building requiring high security inputs from the design stage. Despite some unprecedented designed security the assessment shows not very high level overall due to concentration of security thought to the DC block only and some disproportionate dilutions in the NDC block. The DC block in itself is like a bunker while the front shielding block is not indulged even half as much.

The relevance and importance of site selection as a necessary approach is again underscored here. It can escalate or de-escalate a number of vulnerabilities which may or may not be surmountable in operations.

The research uncovers that this building is far from traditional in its rendition and approach. Yet the simple logic of keeping the critical function shielded works well. At the same time while this exposes the DC block to the non-securable, its structural strengthening makes it a worthwhile shield doing well to screen the human lives. So once again we find that good design approach and effective functional zoning can influence security in a much more comprehensive manner than any amount of mechanical and organised means.

This case study exemplifies how the best of budgets and expertise cannot substitute good planning sense and design. Despite the distinction of being the first blast resistant building in India and very high perceptible security, high resilience and high preventive mechanisms, the actual efficacy is not as high due to some inadvertent lapses and insurmountable vulnerabilities. While some vulnerability has been very skilfully addressed by good common sense design strategy, other design lapses remain unaddressed despite a lot of mechanization and militarization. Here too a lot of the security deployment seems to be feeding the associated perceptions for a National icon of this stature.

4 CONCLUSIONS

The case studies show that the traditional approach following the age-old style and planning systems in design invariably serves well from the security perspective. This may be owing to the high priority that security always received in the olden times using the built space as the primary defence tool. Also the natural materials that were used traditionally whether they were sandstone or wood mostly do not have much potential to splinter or create flying debris. They have greater likelihood to absorb the impact. Also most of the traditional approaches sit in extremely well with the CPTED techniques too. The only drawback is the high land requirement and horizontal sprawl it usually entails that can be a constraint in modern times.

The studies also highlight the importance of site selection as one of the most important aspects to consider to maximise advantage from topography and minimise vulnerabilities. Conventional techniques hinge on the basic common sense of layering and securing the core as the most rudimentary wisdom. While the Akshardham temple displayed its optimum use for high embedded security, the Data centre complex demonstrates the constraints and responses of making do with a given site as a practical contemporary reality.

Another important derivative is the huge disconnect between perceptions and actuality of security. The highly secured 'blast resistant structures' for the Data centre complex, are dealing with some big vulnerabilities at heavy initial and running costs. Likewise the very part of the Akshardham temple which seemed to be the highlight of its security (the high level of screening, scanning and restrictions) is effectively its biggest drawback in an otherwise superior design with subtlety. The tourists who find this militarisation at the entry so intimidating that they don't go beyond, miss out on a campus of extreme tranquillity and spirituality far removed from the worldly fears.

Both the projects were found to be very high on the resilience aspects for their own unique design considerations. This is also owing to the local and National codes that put heavy impetus on designing for natural disasters, fire and life safety. These are mandatory stipulations enforced as pre-requisites for all constructional projects. This exemplifies the effectiveness of local stipulations in creating better environments and better guidance to the designers even if it is per force. At the same time the study also shows the unmindful compliance with security mandate even at the cost of creating insecurities.

The singular greatest achievement of the audit is the dramatic difference in the ambience which comes from reduced mechanized and jarring security and substitution by other means. These may range from structural strengthening (as in the Data centre complex) to the helpful and covert deployment of volunteers to supplement the natural means. In the case of the Akshardham temple these were compulsions of the building function but they exemplify that it can be done for any public domain if the dictat so demands.

ACKNOWLEDGEMENTS

1. Dr. (Prof.) Mandeep Singh, Dean, School of Planning and Architecture and the guide for the Research.
2. Shri Harish Patel, Public Relations officer, Askhardham Temple, New Delhi.
3. Shri P.N. Aggarwal, Retd. DG, Delhi Police and current Security Advisor, Govt. of India.
4. Rajinder Kumar Associates, Architects for the Data Centre complex, New Delhi
5. TPCL Structural engineers for the Data Centre complex, New Delhi.

REFERENCES

[1] Swaminarayan Akshardham, BAPS Swaminarayan Sansthan, New Delhi. Online. www.akshardham.com. Accessed on 7 January 2017 .

[2] Patel, Shri Harish, Personal communication, 18 February 2017, Public relations officer, Akshardham Temple, New Delhi.

[3] Google map of Akshardham temple, New delhi India. Online. www.googlemaps. Accessed on 20 February 2017

[4] Aggarwal, P. N., (Retired Director General of Delhi Police). Personal communication, 27 January 2014, Security advisor to Government of India.

[5] Grover, Manika, Personal communication, 10 September 2016, Project Architect, RKA Architects New Delhi, India.

[6] Singh, Er. Amritpal, Personal communication, 8 September 2016, Structural engineer, TPC engineers, Noida, India

CYBER INCIDENT EXERCISE FOR SAFETY PROTECTION IN CRITICAL INFRASTRUCTURE

YUITAKA OTA, TOMOMI AOYAMA, DAVAAADORJ NYAMBAYAR & ICHIRO KOSHIJIMA
Industrial Management Engineering, Nagoya Institute of Technology, Japan.

ABSTRACT

Many companies, especially those that own critical infrastructure (CI), must prepare processes to cope with serious incidents before they happen. Conventional safety countermeasures already developed a priori to deal with expected problems, such as machinery malfunction, natural disasters and human errors. Field operators also are well trained against such problems. In recent years, however, cyber-attacks have emerged as a 'clear and present danger' and have rendered CI uncertain and unsafe through industrial control systems (ICSs). Thus, CI owners should now prepare countermeasures to ensure the safety and security of ICSs. Unfortunately, responding to situations without experience and developing adequate countermeasures is a difficult challenge. A certain resilience must be developed that gives the actors the ability to flexibly cope with a crisis and quickly recover to a safer state. In CI systems, field operators are the most important element for dynamically managing ICS emergency response.

In this paper, the authors would like to discuss the following two problems:

1. Simultaneous achievement framework of safety and security in ICSs
2. Personnel training methodology based on the above framework

Also, we present an illustrative example of the proposed framework and methods based on exercises in which almost 200 CI personnel and security experts participated.

Keywords: cyber-incident, personnel training, ICS-SIRT

1 INTRODUCTION

1.1 Background

The threat of cyber-attack to industrial control systems (ICSs) has been increasing since Stuxnet was discovered in Iran's uranium enrichment facility in 2010[1]. As a result, companies are obliged to compromise by increasing resilience against such sophisticated cyber-attacks through ICSs. In 2016, a blackout due to a cyber-attack occurred in Ukraine [2]. In this case, restoration was rapidly achieved (about 1 hour) by manual operations. In this case, restoration was rapidly achieved (about 1 hour) by manual operations learned from a previous attack in 2015 [3]. This case suggests that (i) cyber-attacks are a real threat for most ICSs and (ii) the human contribution is essential for responding to a cyber-attack.

It is essential to enhance security to protect companies from cyber-attacks. However, even if companies improve cyber security, preventing all cyber-attacks is impossible. Also, companies holding ICSs, especially Critical Infrastructures (CIs), support our everyday lives. Shutting down for the safety of CIs being threatened by cyber-attacks has a significant impact on social and business activities. In fact, the result of such a shutdown often satisfies the intention of the attackers. Therefore, companies involved with CI must consider not only plant safety but also business continuity.

DOI: 10.2495/SAFE-V8-N2-246-257

1.2 Problem Statement

Companies holding ICSs have taken various countermeasures to ensure and maintain plant safety. These countermeasures are prepared considering malfunction of equipment, natural disasters, human errors, etc. as threatening plant safety. Not only have the factors that have been examined until now, but also cyber-attacks need to be considered as threatening plant safety. However, Japanese companies have not been seriously considering a cyber-attack as a cause of ICS malfunction. If a cyber-attack occurs, it may propagate not only cyber damage through the ICS network but also physical damages through the plant's physical structure. Thus, owners of CI should prepare countermeasures to ensure the safety and security of ICSs.

In Ref. [4], E-ISAC makes the following implications:

'The attacks highlight the need to develop active cyber defenses, capable and well-exercised incident response plans and resilient operations plans to survive a sophisticated attack and restore the system'.

As an example of such an approach, for the 2020 Olympic and Paralympic Games, the Japanese Ministry of Economy, Trade and Industry has promoted the 'Cybersecurity Management Guideline,' which includes human resource development in the industrial sector [5].

2 CYBER-ATTACKS ON INDUSTRIAL CONTROL SYSTEM

Figure 1 shows a cyber security testbed and a virtual ICS network. The testbed (Fig. 1) heats water with a heater in the lower tank and circulates the heated water through a simple pipeline to the upper tank (using a pump). The testbed is also equipped with actual industrial control devices and controlled automatically. In this ICS network, OLE for Process Control (OPC) Servers collect and exchange process data, and monitor them by using Supervisory Control And Data Acquisition (SCADA) function included in the OPC Servers. OPC Servers receive various parameters from the testbed's sensor, and SCADA gives instructions to the control equipment using the data of the parameters. This ICS network is designed by the secure zoning method proposed by Hashimoto [6] to suppress the damage of cyber-attacks.

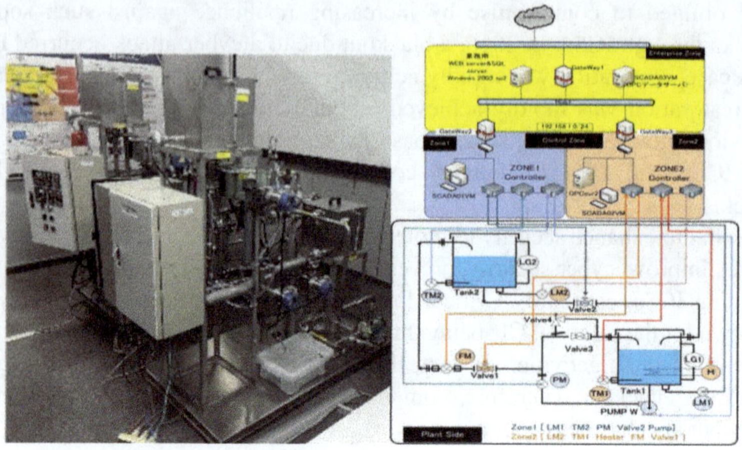

Figure 1: Cyber security testbed and ICS network.

The authors have demonstrated cyber-attacks against the testbed. According to the steps of the anatomy of a cyber-attack [7], the attacking procedure is determined as follows.

1. Attack Step 1: Reconnoitering and enumeration
   The attacker executes the Armitage to launch Metasploit framework [8]. Then, the attacker executes the Nmap, which scans a target network (ICS-2 Network 192.168.12.0/24). After completing the scan, the attacker gets target information.
2. Attack Step 2: Intrusion and advanced attacks
   According to the functions of the Armitage, the attacker ascertains an attack method and vulnerability in the Metasploit framework. The attacker exploits payloads, which execute inappropriate code in the use of the vulnerability. After sending the payloads, the attacker obtains administrator's permission for the OPC2 illegally.
3. Attack Step 3: Malware insertion
   The attacker uploads a malware execution file to an OPC2 background process so as to execute a malware file. The configuration file of OPC2 was rewritten, and then the OPC2 was restarted. Resulting from the attack, the valve controlled by the OPC2 shutdown.
4. Attack Step 4: Clean-up
   The attacker deletes the malware file and the configuration file from OPC2 storage so as not to leave the cyber-attacks evidence, and finish the cyber-attack.

Figure 2 shows monitoring screen shot in operation. As the valve is opened, the stage of the testbed changes. Usually, the state transition of the testbed appears on the monitor of the operator as shown in the left figure. The changing graph shows the liquid level of the upper tank. The liquid level is changing due to malfunction of the valve. The operator can notice the abnormality of the testbed due to the unusual process deviations from the steady state condition through the monitor. However, the attacker conceals the information about cyber-attack in order not to let anyone notice the cyber-attack. The monitoring screen of the testbed is concealed as shown in the right figure, while the cyber-attack causes serious accidents at the testbed.

This demonstration shows a possibility of safety incidents caused by cyber-attacks to ICS.

3 INCIDENT RESOPNSE AND ORGANIZATIONAL STRUCTURE

To date, malfunction of equipment, human error, etc. have been considered as causes of ICS malfunction. In addition to this reason, companies need to add cyber-attacks as a cause of ICS malfunction. Abnormal operation of a plant depends on the characteristics of the plant

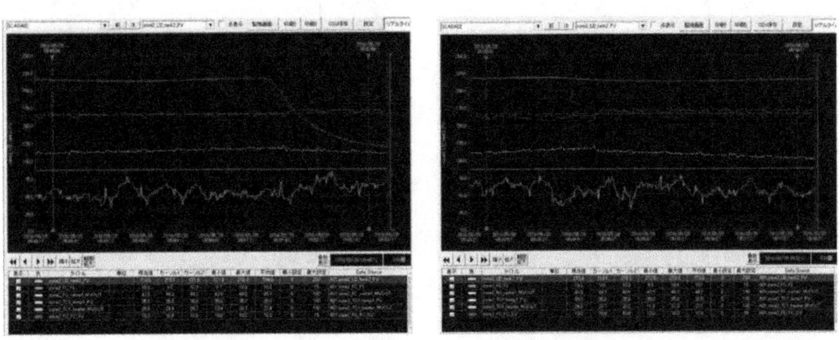

Figure 2: Monitoring screen shot in operation.

regardless of the cause of the ICS malfunction. In other words, the accident that occurs in a plant is the same regardless of whether the cause is a cyber-attack or another reason. Even if it is caused by cyber-attacks, the situation happening at a plant is the same as what has been experienced up to now.

3.1 Incident Response

Companies have made a lot of efforts to ensure the safety of their plants for many years. One of them includes response against plant abnormality. As mentioned previous section, even if plant abnormality caused by cyber-attacks is not different from the situation of the plant that companies have been experiencing until now, they can ensure the safety of the plant threatened by cyber-attacks by using this response (safety response).

Figure 3 shows the structure of cyber incident response. If the plant is operating abnormally caused by the cause considered so far, it is possible to restore plant safety by performing safety response as shown in the left figure. However, there is a possibility that cyber-attacks will interfere with safety response (e.g. concealment of the information of a plant on the monitor). Companies must consider where cyber-attacks interfere safety response. And it is necessary to take the response to eliminate the effect of cyber-attacks (security response) to take safety response as shown in the right figure. Therefore, if a cyber-attack causes abnormal operation of a plant, a response to rectify the situation is required in combination with a structure to add a security response to the safety response.

3.2 Organization Structure Required for Incident Response

In Japanese companies, when an incident happened at a plant, departments and divisions related in places where abnormality is caused have investigated the cause and responded. However, in the case of a cyber-attack (cyber incident), there is a possibility that damage other than the part where the abnormality is happening (infecting other equipment, etc.). In other words, various departments and divisions need to deal with cyber incidents. And if

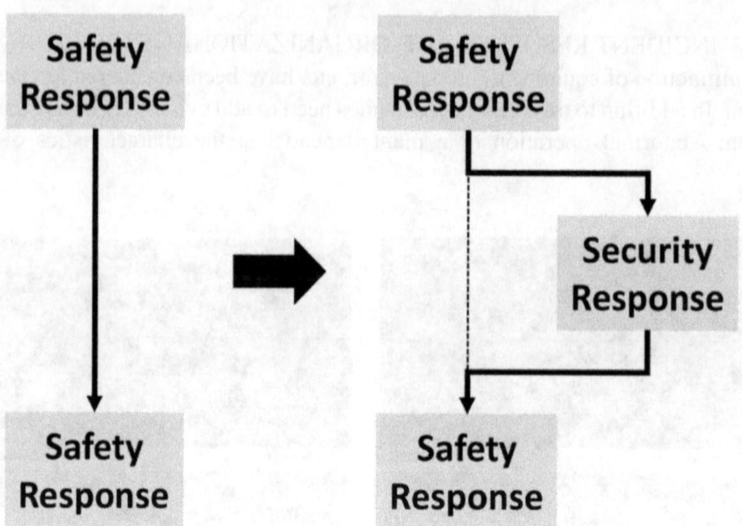

Figure 3: The structure of cyber incident response.

these departments and divisions deal with cyber incidents their own hands, proper response would be delayed, and the damage caused by cyber incidents could be expanded.

To respond proper response, it is necessary for companies to prepare an organizational structure according to the purpose to be achieved beforehand and to dynamically change the organizational structure in accordance with the trend of the situation (change in purpose).

The authors visualized the state in which the work volume changes over time (Fig. 4). And, we visualized the organizational structure required for incident response as well as the transition of the work volume.

a) Operation Staff
The purpose of this organization is to operate the plant normally and provide company's services. The operation Staff is familiar with the normal operation of the plant, so he/she can detect the behaviour of the plant different from normal operation.
b) ICS-ERT (Emergency Response Team)
The purpose of this organization is to ensure the plant safe whatever the cause is. Instead of focusing on the cause of plant malfunction, ICS-ERT will focus on the state of the plant and take response. Moreover, ICS-ERT do not work in herself, but manage the Operation Staff (including the IT Staff in the field) and ensure the safety of the plant.
c) ICS-SIRT (Security Incident Response Team)
The purpose of this organization is to respond to incidents and return to the state of normal plant for continuity Company's business. In the case of a plant abnormality caused by a cyber incident, it is necessary to respond to IT to ensure the safety of ICS not only for ICS.
d) CSIRT (Computer Security Incident Response Team)
The purpose of this organization is to respond to parts (IT system) other than the ICS-SIRT service (ICS system) and return it to the state of an ordinary enterprise IT system.

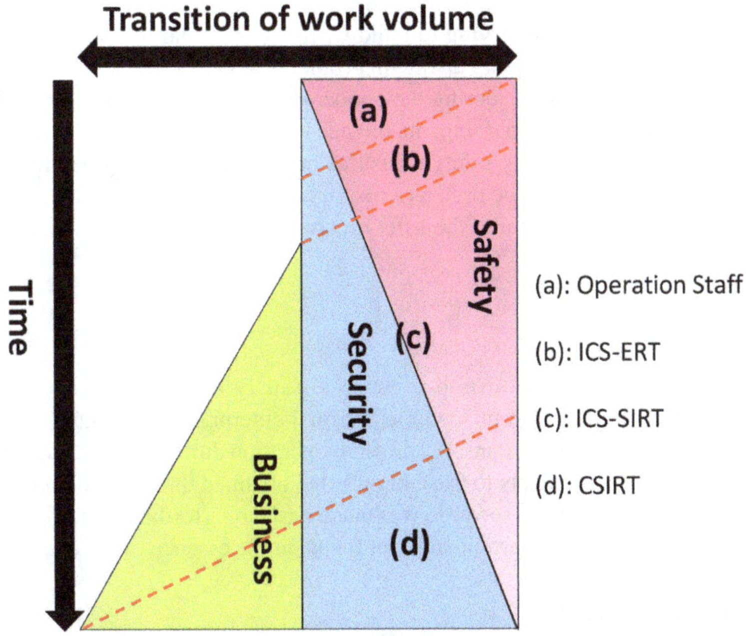

Figure 4: Organization image for cyber incident response.

4 EXERCISE IN BUSINESS CONTINUITY PLANT

To formulate measure to ensure the safety and security of ICS, it is necessary to experience how cyber-attacks affect ICS. By getting those experiences, companies can understand what measure can be taken to remove the effect of cyber-attacks. However, many companies don't have the experience of cyber-attacks to formulate measure against cyber-attacks (companies don't have the ability to detect even if it is cyber-attack). Therefore, to think about the measure of the inexperienced situation, the exercise, which can simulate the cyber-attack, can be effective.

4.1 Purpose of the Exercise

The purpose of the exercise is to 'learn how to think about organizational collaboration to deal with cyber-attacks on ICS.' Therefore, the object of the exercise is as follows;

Organization:	Organization concerned with administration of ICS Organization concerned with administration of business
Response:	Safety management (safety response), response to cyber-attack (security response)

Given the combination of organization and response, exercises must satisfy the following three requirements for participants:

- Appropriate information must be gathered for those who determine the source of the attack. If necessary information cannot be collected, the person determining the source of the attack cannot formulate an appropriate response. As a result, the response may be delayed, and the requisite information. In other words, communication must be smooth. Therefore, the exercise must enable participants to understand the importance of communication when dealing with abnormal situations.
- Many companies in Japan are trying to build CSIRT to respond to cyber incidents. However, CSIRT for IT systems cannot deal with problems related to ICSs. ICS-SIRT and ICS-SIRT are necessary when considering the cooperation of the targeted organization. Thus, the exercise makes it clear to participants the need for ICS-ERT and ICS-SIRT.
- Many companies are preparing safety measures in case of abnormal operation of a plant. This safety response can also be used as a response to cyber-attacks. Thus, participants in the exercise must understand that the safety response is available in the case of a cyber-attack.

4.2 Design of Exercise

It is effective to construct a case-based exercise to satisfy the requirements as mentioned above. The exercise consists of four scenarios: virtual enterprise, cyber-attacks, response to operational situation that becomes unsafe and plant operation during cyber-attack. It is needed to make it possible for participants to use a knowledge obtained through the exercise for their company. Therefore, the exercise scenario was made based on a testbed (Fig. 1). The testbed's structure is kept simple so that participants can focus on the essence of the exercise.

- Virtual enterprise scenario
  In this scenario, participants must think about how organizations can cooperate to handle cyber-attacks. To make it easier for participants to imagine organizational collaboration,

an organizational structure equivalent to that of the real company must be set up to the extent possible.

- Cyber-attacks
  This scenario requires participants to think about how to respond to incidents caused by cyber-attacks. Participants should be as real as possible so that participants gain experience in dealing with realistic cyber-attacks.
  This scenario is created based on the cyber-attack the authors are conducting.
- Response to operational situation that becomes unsafe
  This scenario allows participants to understand how cyber-attacks affect the safety response.
  This scenario simulates an abnormal operation of the testbed. The safety response to the situation is written in the workflow (Fig. 5).
- Plant operation during cyber-attack
  This scenario allows participants to understand the full impact of ICSs malfunction caused by a cyber-attack.
  This Scenario is created by observing the transition of the driving situation of the testbed due to the cyber-attack the authors are conducting.

4.3 Administration of Exercise

During the proposed exercise, participants should think about how it corresponds to a situation caused by a real cyber-attack and how this correspondence evolves over time. In the exercise, participants think about the reaction to the situation caused by the cyber-attack. If there are few coping people, the coping person cannot cope with the lack of resource. The reaction is thereby late. If the response is delayed, the attacker leads. If the attacker leads, the damage spreads. To prevent the damage spreads, it is necessary for a person coping with the

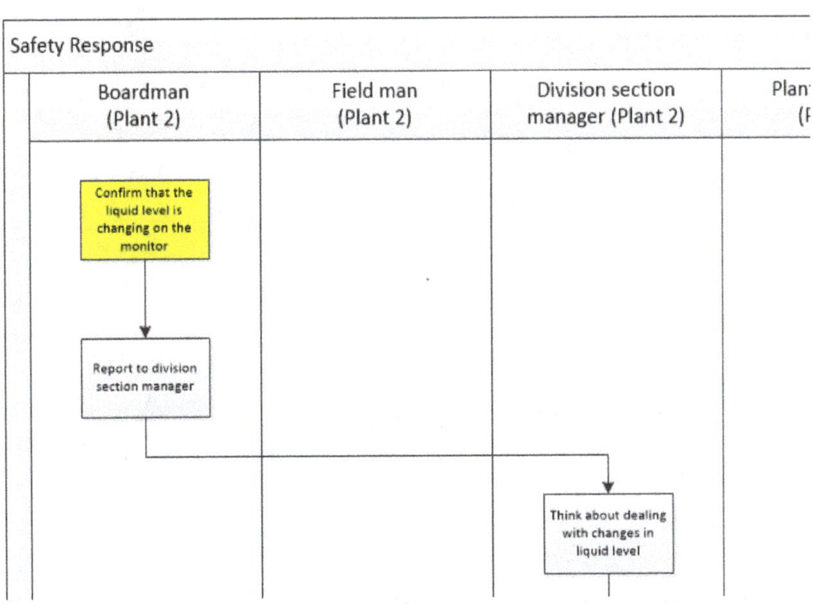

Figure 5: Safety response in the workflow (part).

situation caused by the cyber-attack to reconstitute an organization dynamically to avoid the attacker from leading.

To make effective responses, participants have to understand attacker's scenario. To ensure that participants understand the attacker's intention, the exercise is divided into three phases according to the attacker's scenario (Fig. 6).

1. Predictive phase
   The purpose of this phase is to ensure the safety of plants threatened by cyber-attacks. In this phase, the participants examine the safety response obstructed by a cyber-attack.
   After detecting suspicious communication in a section that monitors the network, the exercise begins with an alert for the status of the liquid in the plant. In this phase, the participants must ensure the safety of the plant. Once safety is ensured, participants must investigate the abnormal operation. In this phase, participants can recognize how different causes can influence the safety response.
2. Emergency phase
   The purpose of this phase is to continue business. The participants consider about the security of the plant and how the exercise implicates to a real cyber-attack. They must also take into account how their business will be affected by the cyber-attack. In this phase, the participants consider practical measures taken to ensure business continuity throughout the company.
   This phase begins after the abnormal operation of the plant reveals a cyber-attack. In this phase, participants can recognize that the entire company must be mobilized to treat a situation caused by a cyber-attack.

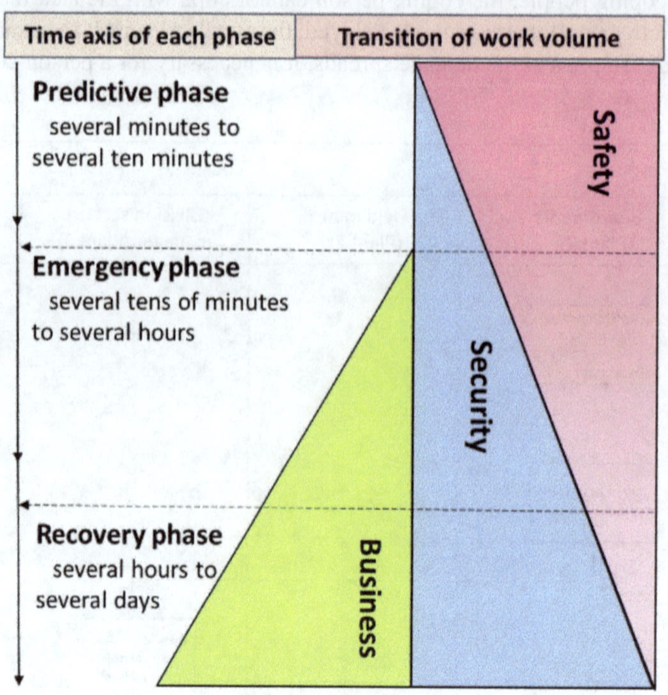

Figure 6: The structure of exercise with workload image for safety, security and business continuity activities.

3. Recovery phase
   The purpose of this phase is to restore own business to normal operation rapidly. The participants must find a way to nullify the influence of the cyber-attack so that the plant can return to normal operation. In addition, they must consider how business is affected by the temporary plant shut down. In this phase, the participants examine the start-up procedure of a plant crippled by a cyber-attack. In addition, the participants examine methods to prevent the recurrence of similar cyber-attacks.
   This phase begins after the plant is stopped by a cyber-attack. In this phase, participants think about how to restart the plant. In this phase, participants can see how different causes affect the restoration of normal plant operation.

4.4 Administration Method of Exercise

This exercise is done in groups that consist of four to six people. To consider organizational correspondence, participants come from different departments. The exercise requires participants to think about systematic countermeasures and collaborative communication. Therefore, no specific role is given to group members.

4.4.1 Exercise method

The exercise is conducted among members who have different backgrounds. The exercise is a discussion-based exercise for discussion will be active, compare participant's views on a common problem (cyber security) and achieve the purpose of the exercise.

4.4.2 Deliverables of Exercise

The typical deliverables of the exercise are workflow diagrams (Fig. 7), which become plans by considering 'who' performs the action and 'when.' Therefore, the horizontal axis of the worksheet corresponds to the department that does the action and the vertical axis of the worksheet corresponds to the timeline.

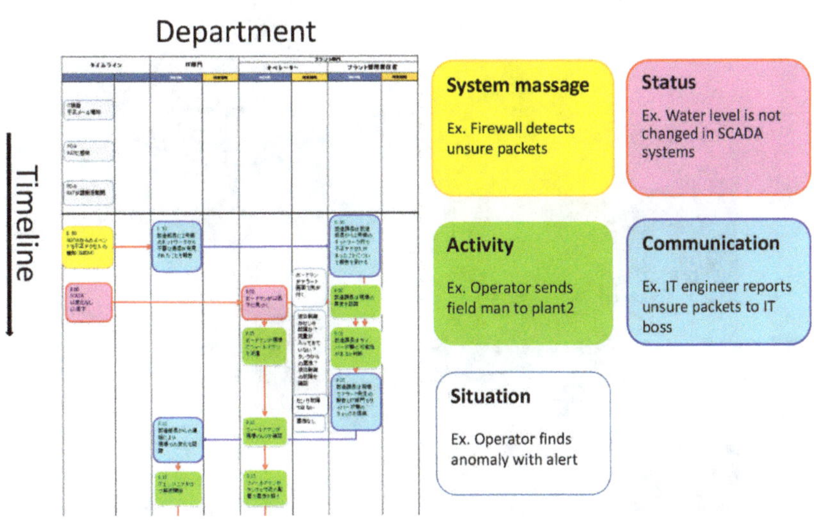

Figure 7: Sample workflow diagram.

4.4.3 Implementing Method

This exercise consists of five steps: an introduction, explanation of preconditions, group work, presentation of results and open discussion. As mentioned in the previous section, the exercise is divided into three phases. Of the five steps, explanation of preconditions, group work and presentation of results occur in each phase as one cycle.

1. Introduction
   The facilitator explains the purpose of the exercise and how to proceed. The general scenarios of the exercise are introduced to the members. At this step, members understand the importance of the exercise and should, therefore, be motivated to apply themselves fully in the exercise. By doing so, members can increase the benefit of the exercise.
2. Explanation of Preconditions
   Preconditions are set for each phase. Preconditions include the state of the plant, the state of the IT network, and the company situation. At the beginning of each phase, the facilitator explains the preconditions. At this point, members watch a video to understand the preconditions (the authors created the video and reproduced the preconditions).
3. Group Work
   The group work consists of workflows that add a security response to the safety response. An A0-size worksheet, colored sticky notes, and a marker are distributed to each group. A precondition and safe response are printed on each worksheet. After understanding the precondition and safety response written in the worksheet, the members do the group work. In group work, if a safety response is deemed insufficient, members add a response by writing the response on a sticky note and sticking it to the worksheet. The meaning of each sticky note depends on its color, so members should use sticky notes of the same color to express similar activities. Next, consider the relationship between the added response and the safety response and draw a line with a marker between the two. Through this exercise, members can visualize the response to cyber-attacks and the structure of the organization collaboration. Members can debate about the various departments involved based on the worksheet (Fig. 8).

Figure 8: An example of team discussion in group work.

4. Presentation
   Each group presents their work flow created in the group work while displaying their worksheet to all participants. In this presentation, each group explains not only the added responses but also why they were added.
5. Open-Discussion
   Members share among themselves the knowledge gained through the exercise. Sharing the knowledge through discussion increases the benefit of the exercise for all participants. Also, the administration and the members discuss whether the exercise was useful and whether the administration of the exercise was appropriate.

4.4.4 Administration Staff

The participants in this exercise are supposed to be about 30 to 40 members. The size and complexity of this exercise require a lot of help from the person to progress the exercise smooth. However, even if there are many people, the exercise does not proceed smoothly. For a smooth administration of the exercise, the role of administration staff divided as follows: facilitator, advisor and supporter.

a) Facilitator: The facilitator guides participants through the exercise. He/she explains purpose and method of the exercise at the introduction and the precondition at each phase. During the group work, the facilitator pays attention to each group's progress and manage the time of group work. After the group work is finished, the facilitator will ask questions of each group's deliverables so that the discussion will become active.
b) Adviser: the advisor participates in each group and supports members. The advisors don't actively support, but instead answer questions raised in the group work, explain preconditions in detail and provide advice on creating the workflow.
c) Supporter: The supporter is not directly involved in the exercise. However, it's hard to exercise without this role. The supporter sets up the exercise meeting place before exercise begins and distributes necessary tools for the group work.

4.5 Trial Exercise

To date, a total of 200 professionals has participated this exercise. They expressed a mostly positive opinion about the exercise method and the administration method. Figure 9 shows the level of satisfaction of the participants. According to this figure, the members were mostly satisfied with the exercise.

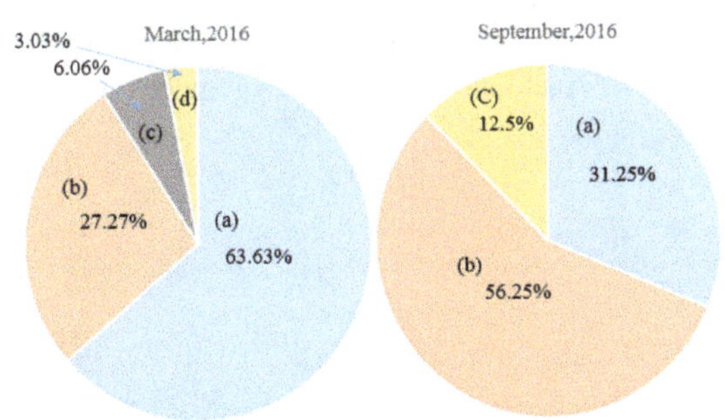

Figure 9: Satisfaction of exercise participant.

Participants understood the purpose of the exercise and recognized its importance. However, this exercise differs in many ways on a real company. Thus, many problems can be expected if this exercise is applied to a real company.

5 CONCLUDING REMARKS

In this paper, the authors explain the incident response and organization for incident response for the safety and security of ICS and examine exercises developed to train employees how to respond in the case of cyber-attacks.

In the future plan, by analyzing the deliverables obtained in the trial exercises, we will be able to improve the exercise. We derive the organizational structure and communication method for dynamical transformation of organizational structures. And we will develop a generic exercise package to be customized to extend this ICS-BCP exercise to various CI companies.

ACKNOWLEDGEMENTS

This research is partially supported by the Ministry of Education, Science, Sports and Culture, Grant-in-Aid for Scientific Research (A), No.16H01837 (2016) and Council for Science, Technology and Innovation (CSTI), Cross-ministerial Strategic Innovation Promotion Program (SIP), 'Cyber- Security for Critical Infrastructure' (Funding Agency: NEDO), however, all remaining errors are attributable to the authors.

REFERENCES

[1] Symantec Stuxnet – Modus Operandi. Available at: https://www.sans.org/summit-archives/file/summit-archive-1493844778.pdf. (Accessed July 4, 2017).

[2] The Center for Strategic Cyberspace + Security Science. Available at: http://cscss.org/CS/2017/01/19/ukraine-confirms-december-kiev-blackout-was-cyber-sabotage/. (Accessed July 4, 2017).

[3] SANS Industrial Control Systems Security Blog. Available at: https://ics.sans.org/blog/2016/01/09/confirmation-of-a-coordinated-attack-on-the-ukrainian-power-grid. (Accessed July 4, 2017).

[4] E-ISAC, Analysis of the Cyber Attack on the Ukrainian Power Grid (March 18, 2016).

[5] Japanese Ministry of Economy, Trade and Industry. Available at: http://www.meti.go.jp/english/press/2015/1228_03.html. (Accessed July 4, 2017).

[6] Hashimoto, Y., Toyoshima, T., Yogo, S., Koike, M., Hamaguchi, T., Jing, S. & Koshijima, I., Safety securing approach against cyber-attacks for process control system. *Computers & Chemical Engineering*, **57**(15), pp. 181–186, 2013. https://doi.org/10.1016/j.compchemeng.2013.04.019

[7] Dell Sonicwall, Anatomy of a cyber-attack, Dell software, 2012.

[8] Metasploit Framework. Available at: http://www.metasploit.com/. Accessed July 4, 2017).

IDENTIFICATION OF POTENTIAL HAZARDOUS EVENTS OF UNLOADING SYSTEM AND CO_2 STORAGE TANKS OF AN INTERMEDIATE STORAGE TERMINAL FOR THE KOREA CLEAN CARBON STORAGE PROJECT 2025

HYONJEONG NOH[1], KWANGU KANG[1,*], CHEOL HUH[2], SEONG-GIL KANG[1] & YOUNGKYUN SEO[1]
[1] Korea Research Institute of Ships & Ocean Engineering, Daejeon, Republic of Korea.
[2] Ocean Science & Technology School, Korea Maritime and Ocean University, Busan, Republic of Korea.

ABSTRACT

Carbon capture and storage (CCS) is regarded as one of key technologies to meet global greenhouse gas emission reduction goal. In this manner, South Korea is developing a one-million-ton-scale offshore CCS project to reduce national greenhouse gas emission, entitled the Korea Clean Carbon Storage Project (KCCS) 2025. In this project, the CO_2 is captured from power plants on coast, loaded to CO_2 carriers, and transported to an intermediate storage terminal, which is located on shoreline nearby offshore storage reservoirs at the Ulleung Basin. Then the CO_2 is exported via an offshore pipeline to offshore platform for injection to geological storage site for a permanent containment. Since the concept of the KCCS 2025 includes CO_2 carriers as a transportation method, it requires an intermediate storage terminal that receives the CO_2 from carrier and send out continuous CO_2 flow to the offshore pipeline after the pressurization of CO_2 to higher than 100 bar. The intermediate storage terminal will consist of (1) unloading system, (2) CO_2 storage tanks, (3) LP pumps, (4) a reliquefaction package, (5) a vent stack and (6) HP pump and injection pump (a booster station). Because there are few actual projects with an intermediate storage terminal worldwide, researches on intermediate storage facilities are insufficient. To support an optimal concept design of the intermediate storage terminal for the KCCS 2025, this study identified the potential hazardous events for the unloading system and the CO_2 storage tanks. For the unloading system, an unloading arm and LCO_2 recirculating line are found to be major components causing serious damage in case of accident. In the case of the CO_2 storage tanks, where large amounts of CO_2 are stored, they can cause serious damage due to large amounts of CO_2 leakage when the tanks are ruptured, over-pressurized, low-pressurized, overcharged, etc. Because hazardous events may pose significant harm to humans or the environment, these results should be considered in the next phases of the project. The results of this study can be helpful for the development of safe CO_2 transportation technology in the future.

Keywords: CCS, CO_2 storage tank, CO_2 storage terminal, hazard, PHA, unloading arm

1 INTRODUCTION

Worldwide efforts are being made to reduce GHG emissions in response to climate change. As a bridge technology, the importance of carbon dioxide capture and storage (CCS) is emphasized, which captures CO_2 from a large emission sources such as thermal power plants, and permanently stores in sediments between 800 and 3,000 m underground [1]. In Korea, the industrial structure is focused on manufacturing, and therefore, it is necessary to make structural changes in the industry to reduce CO_2 emissions in Korea. Various practical measures are being taken by the Korean government. As part of these efforts, the Korean government is preparing a 1 million-ton offshore CCS project [2].The title of this 1 million-ton offshore CCS project, which is currently in the process of the conceptual design, is Korea Clean Carbon Storage Project (KCCS) 2025. In the KCCS 2025 project, CO_2 is captured from a thermal power plant located on the west coast of Korea and transported to an intermediate storage terminal via CO_2 carriers. Subsequently, the CO_2 is transported to a platform via an offshore pipeline, and then injected to reservoirs located in the Ulleung Basin. What is

DOI: 10.2495/SAFE-V8-N2-258-265

notable about KCCS 2025 is that it uses CO_2 carriers as a transport method from the CO_2 source to the intermediate storage terminal. Because it is very difficult to inject the CO_2 continuously to offshore geological storage site directly from CO_2 carrier, the CO_2 transport method using carriers usually requires a CO_2 intermediate storage terminal. However, there are few studies on a CO_2 intermediate storage terminal, especially on the risk analysis. Based on the conceptual design of KCCS 2025, in this study, preliminary hazard analysis (PHA) is conducted on two facilities of the intermediate storage terminal, the unloading system and the CO_2 storage tanks. The PHA results of this study can be important on the reliability and risk analysis in the FEED study of a CO_2 intermediate storage terminal [3].

2 CO_2 INTERMEDIATE STORAGE TERMINAL

The targets of this study are the unloading system and the CO_2 storage tanks of the CO_2 intermediate storage terminal of the KCCS 2025 project. The CO_2 intermediate storage terminal aims for temporary storage of liquid CO_2 transported through 8K CO_2 carriers. After pressurization and heating process at the terminal, the CO_2 is sent to the offshore platform for injection through the offshore pipeline.

The KCCS project uses three 8K CO_2 carriers to transport liquefied CO_2 from the source to the intermediate storage terminal. The schedule of CO_2 carriers can vary depending on the weather, and the terminal should also act as a buffer station of CO_2 transportation. The temperature and pressure conditions of the CO_2 carrier cargo tank and the CO_2 storage tanks are the same at 1.5 Mpag and -27 °C. The CO_2 intermediate storage terminal consists of six facilities as shown below.

1. Unloading system (including a vapor return arm)
2. CO_2 storage tanks
3. LP pumps
4. Reliquefaction package (including a pressure built-up vaporizer)
5. Vent stack
6. HP pump & injection pump (a booster station)

The unloading system works for liquid CO_2 unloading and vapor CO_2 loading. The CO_2 storage tanks store the CO_2 transported from the CO_2 carriers. The LP pumps act as a pressurizer to send CO_2 in the storage tank to the booster station. Reliquefaction package re-liquefies boiled-off vapor CO_2. The pressure-built up vaporizer produces vapor CO_2 to prevent negative pressure in the CO_2 storage tank. The vent stack serves to release the vapor CO_2 generated in the emergency to the atmosphere. The HP pump and the injection pump are responsible for pressurizing the LCO_2 pressure to around 100 bar, which is suitable for CO_2 injection. In this study, the hazard of (1) the unloading system and (2) the CO_2 storage tanks are analysed, which are considered to have high risks among the above six facilities. The unloading system and the CO_2 storage tanks are described in more detail below.

2.1 Unloading system

When the CO_2 carriers arrive at the CO_2 intermediate storage terminal located in Ulsan port, LCO_2 is unloaded using an unloading arm with 2000 m^3/h rate. The LCO_2 is transported to the CO_2 storage tanks through 24-inch unloading pipeline. The pressure of CO_2 emitted from the CO_2 carrier is designed as 2.1 Mpag considering the pressure reduction in the unloading arm and the unloading pipeline, and the height and storage condition of the CO_2 storage tanks. When LCO_2 is unloaded from the CO_2 carrier, the cargo tank in the carrier should be charged with the vapor CO_2 to prevent negative pressure in the cargo tank. For this purpose, a vapor CO_2 return arm and a vapor CO_2 return line should be installed. To maintain the temperature of the CO_2 unloading line, the recirculation line should be installed and some of the LCO2 separated from the HP Pump should be flowed.

The unloading system is divided into six components and PHA is performed on each component. The first component is the 14-inch unloading arm, which is responsible for unloading the LCO_2 transported through the CO_2 carrier. The second component is the vapor CO_2 return arm, which is responsible for loading vapor CO_2 into the CO_2 carrier to compensate for the negative pressure in the cargo tank. The third component is the 24-inch unloading line, which is the main pipeline connecting from the unloading arm to the CO_2 storage tank. The fourth component is the 8-inch recirculation line that circulates some LCO_2 separated from the booster station to keep the unloading line at a low temperature. The fifth component is a 10-inch vapor CO_2 return line that supplies vapor CO_2 to prevent negative pressure on the cargo tank of the CO_2 carrier. The 6th component is on-off valves.

2.2 CO_2 storage tanks

The CO_2 storage tanks consist of four 5,000 m^3 spherical tanks. The 8K CO_2 transported by the CO_2 carrier is stored in two 4K storage tanks, respectively. At the same time, vapor CO_2 in the tank is sent to the cargo tank of the carrier to prevent the negative pressure. LCO_2 in the one of the rest two tanks is transmitted through the LP pumps to the boosting station, and the other tank acts as a buffer tank. During the CO_2 transmission, negative pressure can be generated in the tank. It is necessary to charge vapor CO_2 from the vaporizer to the tank to prevent any failure of the tank induced by negative pressure.

The CO_2 storage tanks are divided into five components as follows. The first component is the storage tank itself. The second component is the 24-inch LCO_2 charging line that transports the LCO_2 from the unloading system to the CO_2 storage tanks. The third component is the 6-inch LCO_2 transmission line connecting the storage tank to the LP pumps. The fourth component is the 2-inch pressure built-up vapor CO_2 line to prevent the negative pressure in the storage tanks when LCO_2 is transmitted to the LP pumps. The fifth component is the 3-inch vapor CO_2 release line that discharges vapor CO_2 from the storage tanks to the cargo tank in the carrier or reliquefaction package.

3 PROCEDURE OF PHA

In this study, PHA is conducted for the unloading system and the CO_2 storage tanks of the CO_2 intermediate storage terminal. The purpose of the PHA is to identify the potential hazardous events of facilities at the conceptual design stage and to proceed FFED study taking into account the PHA results [4]. The results can be used for the reliability and risk analysis of CO_2 intermediate storage facilities in the future. The unloading system is divided into 6 components and the CO_2 storage tanks are divided into five components to perform the PHA. Subsequently, potential hazardous events are then derived for each component. After that, cause and effects, and the risk-reducing measures are derived for each event.

4 RESULTS

4.1 Unloading system

The unloading system consists of six components as shown in Fig. 1. Table 1 summarizes the PHA results of potential hazardous events and risk-reducing measures for each event in the unloading system. The six components can be divided into three categories depending on risk characteristics as follows.

The first category is the loading/unloading arms. Rupture in the arm or connection failure of the arm with the carrier can cause low temperature vapor and solid CO_2 leakage and may threaten the safety of the operator. In particular, the poor connection between the 14-inch LCO_2 unloading arm and the carrier is very dangerous. It is also necessary to prevent additional leakage of CO_2. Moreover, the arm or the manifold of the carrier may be damaged or CO_2 may leaks from the unloading arm if it is out of the operating envelope due to excessive movement of the carrier by heavy weather condition. In this situation, the arm should be disconnected from the manifold of the carrier safely. In order to cope with the potential hazard that may arise in the arm, it is necessary to install an emergency shutdown valve (ESDV) to isolate the damaged areas, and to install the emergency release system (ERS) to protect the arm or manifold.

The second category is the pipelines of the unloading system. The potential hazardous events in this category include (1) CO_2 leakage due to rupture in the LCO_2 unloading line, the vapor CO_2 return line, and the recirculation line, (2) CO_2 flow rate out of normal range, or (3) pressure increase due to insulation failure. It is necessary to install an emergency shutdown device such as ESDV in case of CO_2 leakage due to rupture in lines. Sufficient temperature

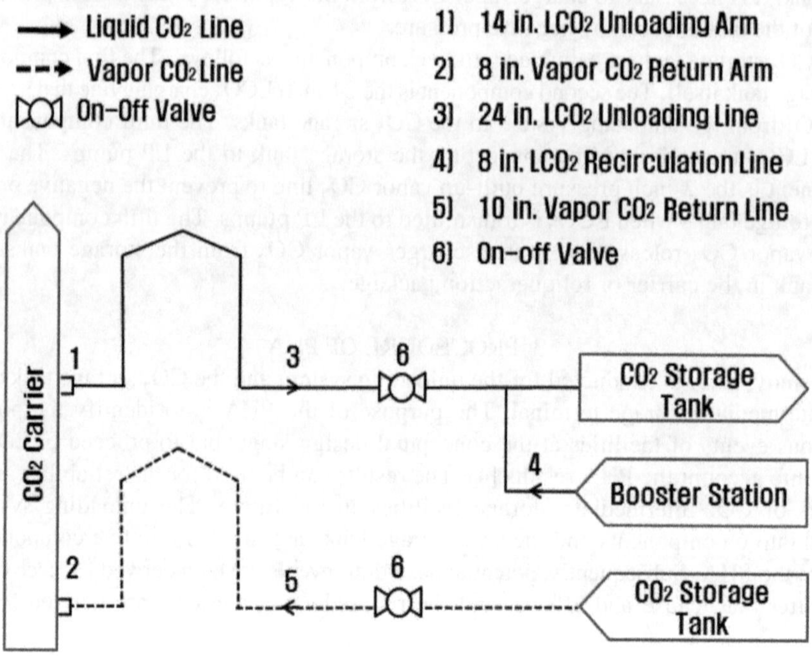

Figure 1: Schematic diagram of unloading arm.

Table 1: PHA results of the unloading system.

No.	Components	Potential hazardous events	Effects	Risk-reducing measures
1	14-inch LCO_2 unloading arm	Rupture or leakage	• Low temperature vapor and solid CO_2 leakage, and may harm workers	• Activate ERS to safely separate the arm from the CO_2 carrier • Stop the pump on the CO_2 carrier • Activate ESDV
		Bad connection	• Low temperature vapor and solid CO_2 leakage, and may harm workers	
		Excessive the CO_2 carrier movement	• Low temperature vapor and solid CO_2 leakage, and may harm workers • Damage of the unloading arm	
2	8-inch vapor CO_2 return arm	Rupture or leakage	• Low temperature vapor CO_2 leakage, and may harm workers	• Activate ERS to safely separate the arm from the CO_2 carrier • Activate ESDV
		Bad connection	• Low temperature vapor CO_2 leakage, and may harm workers	
		Excessive the CO_2 carrier movement	• Low temperature vapor CO_2 leakage, and may harm workers • Damage of the vapor return arm	
3	24-inch LCO_2 unloading line	CO_2 surge	• Increase pipeline rupture possibility due to pressure rise	• Stop the pump on the CO_2 carrier spot • Install PSV
		Rupture or leakage	• CO_2 storage tank is depressurized • CO_2 is released around the line, and pressure drop in the line • CO_2 backflow from LCO_2 storage tank	• Activate ESDV and ERS • Close the valve before and after the damaged location to isolate
4	8-inch LCO_2 recirculation line	Rupture or leakage	• Decrease in cooling efficiency due to CO_2 leakage	• Activate ESDV when low pressure detected
		Temperature increase due to insulation failure etc.	• Temperature increase causes malfunction of entire unloading system	• Activate ESDV when high pressure detected • Install TSV
5	10-inch vapor CO_2 return line	Rupture or leakage	• Vapor CO_2 leakage • Negative pressure and temperature drop in the CO_2 carrier cargo tank	• Activate ESDV (and ERS) • Close the valve before and after the damaged location to isolate
		Increase vapor CO_2 flow	• Increase pipeline rupture possibility due to pressure rise	• CO_2 release through a vent stack
6	On-off valve	Failure to open	• CO_2 unloading is impossible • Increases the unloading system rupture possibility due to pressure increase	• Activate ESDV
		Failure to close	• CO_2 Leakage • CO_2 backflow from the LCO_2 storage tank	

safety valve (TSV) or pressure safety valve (PSV) must be installed in order to prevent damage in lines by abnormal CO_2 flow rate or increased temperature. In particular, the insulation failure in the 8-inch LCO_2 recirculation line is very dangerous because it may cause the overall temperature increase of whole unloading system. Therefore, the risk of recirculation line should be sufficiently reduced by installing TSVs.

The third category is the CO_2 leakage or pressure increase due to the failure of opening and closing of the on-off valves. To prevent these hazardous situations, an emergency shutdown device such as ESDV should be installed and the valves should be properly managed.

4.2 CO_2 storage tanks

The CO_2 storage tanks consist of five components, as represented in Fig. 2. Table 2 summarizes the PHA results of potential hazardous events and risk-reducing measures for each event in the CO_2 storage tanks. The 5 components can be divided into two categories depending on risk characteristics as follows.

The first category the storage tanks them. The main hazardous situations of the storage tanks may be overpressure / low pressure / overcharge due to operational error or leakage caused by rupture. If rupture or leakage occur in the CO_2 storage tanks due to various situation such as drop object, about 4K CO_2 may be leaked, which could lead to disastrous accidents. Therefore, the material and thickness of the tanks should be properly designed so that it will not be affected by various accidents. If overpressure or overcharging situations occur, it can also result in a major accident. PSV valves, level alarms must be installed to manage this situation. Also, ESDV should be installed to stop the CO_2 charging process. Low pressure in the tank can also lead to very dangerous situations. In this case, the low-pressure in the CO_2 tank should be managed by supplying vapor CO_2 through the pressure built-up vapor line.

The second category is pipelines attached to the CO_2 storage tanks. The possibility of rupture of pipelines is relatively small, but it can cause CO_2 leakage or low pressure in the CO_2 storage tanks. In order to manage this situation, it is necessary to install a low-temperature sensor around pipelines to detect the leak and install ESDV to isolate the rupture area in the pipeline.

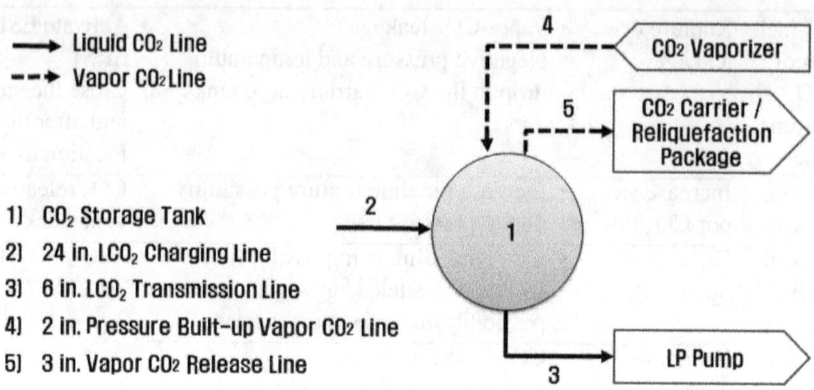

Figure 2: Schematic diagram of the CO_2 storage tanks.

Table 2: PHA results of the CO_2 storage tanks.

No.	Components	Potential hazardous events	Effects	Risk-reducing measures
1	CO_2 storage tank	Rupture or leakage	• CO_2 is released around the storage tank, and can affect the nearby tanks	• Ensure that the material and thickness of the tank are safely designed
		Overpressure	• If the pressure exceeds the design pressure of the storage tank, rupture or leakage can occur at the vulnerable part	• Install PSV
		Low pressure	• Rupture or leakage can occur at the vulnerable part	• Charge vapor CO_2 through the pressure built-up vapor CO_2 line from the vaporizer
		Overcharge	• CO_2 is released around the storage tank, and can affect the nearby tanks	• If the level is above 95%, the alarm sounds. • If the level is above 98%, close the valve(or ESDV) in the LCO_2 charging line
		Turnover	• Pressure increase in the storage tanks	• Safety design needed considering the turnover situation
2	24-inch LCO_2 charging line	Rupture or leakage	• Pressure drop in the CO_2 storage tank • CO_2 is released around the line	• Install a low-temperature sensor around the pipeline to detect CO_2 leakage. • Install ESDV as close as possible to the storage tank
3	6-inch LCO_2 transmission line	Rupture or leakage	• CO_2 storage tank is depressurized • CO_2 is released around the line	• Install a low-temperature sensor around the pipeline to detect CO_2 leakage. • Install ESDV as close as possible to the storage tank
4	2-inch pressure built-up vapor CO_2 Line	Rupture or leakage	• CO_2 storage tank is depressurized • CO_2 is released around the line	• Install a low-temperature sensor around the pipeline to detect CO_2 leakage. • Stop LCO_2 transmission to LP pumps • Install ESDV as close as possible to the vaporizer • Install check valve as close as possible to the storage tank

(Continued)

5	3-inch vapor CO_2 release line	Rupture or leakage	• CO_2 storage tank is depressurized • CO_2 is released around the line	• Install a low-temperature sensor around the pipeline to detect CO_2 leakage. • Close the valve(or ESDV) in the LCO_2 charging line • Install ESDV as close as possible to the storage tank

5 CONCLUSION

In this study, hazard analysis is carried out for the unloading system and the CO_2 storage tanks, which are considered the most dangerous facilities in the CO_2 intermediate storage terminal for offshore CCS transportation process. As a risk analysis method, PHA is used. The unloading system is divided into six components and the CO_2 storage tanks into five components, and it is identified potential hazard events for each components. The effects and risk-reducing measures are derived for each potential hazardous event. For the unloading system, the unloading arm and the LCO_2 recirculating line are found to be dangerous components in case of accidents. In the case of the unloading arm, if an accidents such as a rupture or bad connection with the CO_2 carrier occur, low-temperature solid and vapor CO_2 will leak, which may cause fatal damage to workers. Therefore, installations of ERS, ESDV, etc. are essential. When the LCO_2 recirculating line fails, the temperature of the entire CO_2 intermediate storage facilities can increase, which can cause a serious damage to assets. Therefore, enough TSV should be installed to maintain the low temperature in LCO_2 recirculating line. In the case of CO_2 storage tanks, if accidents such as a rupture, an overpressure, a low pressure, and an overcharge occur, they can cause great damage due to large amount of CO_2 leakage, they can cause great damage due to large amount of CO_2 leakage. Therefore, PSV, level gauge and alarm, and ESDV should be installed to properly manage the dangerous situation. The CO_2 intermediate storage terminal has not yet been installed and operated in the world. However, it is very likely to be introduced if CCS technology is commercialized in the future. Therefore, PHA of this study can be a good reference to develop safe CO_2 transportation technology.

ACKNOWLEDGEMENTS

This research is supported by a grant from Endowment Project of "Acquisition and Application of Reliability Assessment Key-Technologies for Offshore Plant Package/Module" funded by the Korea Research Institute of Ships & Ocean engineering (PES8930).

REFERENCES

[1] IEA, *Energy technology perspectives 2015*, 2015.

[2] Ministry of Maritime Affairs & Fisheries, *Development of Technology for CO2 Offshore Geological Storage*, 2016 (in Korean).

[3] Noh, H., Kang, K., Kang, S.G. & Lee J.G., Strategy for development of HSE management framework for offshore CCS project in Korea. *Journal of Korean Society for Marine Environment and Energy*, **20**(1), pp. 26–36, 2017 (in Korean). https://doi.org/10.7846/jkosmee.2017.20.1.26

[4] Rausand, M., *Risk assessment: theory, methods, and applications*, John Wiley & Sons, Inc., Hoboken, NJ, 2011.

A REAL-TIME, MULTI-SPACE INCIDENT DETECTION SYSTEM FOR INDOOR ENVIRONMENTS

DIMITRA TRIANTAFYLLOU, STELIOS KRINIDIS,
DIMOSTHENHS IOANNIDIS & DIMITRIOS TZOVARAS
Information Technology Institute, Centre for Research and Technology Hellas, Greece.

ABSTRACT

A real-time, multi-camera multi-space incident detection system for indoor environments is proposed. The system focuses on industrial environments and recognizes incidents such as human falls, intrusions to restricted areas and collisions. Building Information Models (BIM) provide all the necessary information for the set up of the system. Furthermore, once an incident is detected an alarm notifying of its type and exact location on the building is triggered so that the appropriate coping mechanism can hasten promptly to the point of the incident. Based on building information about each area particularities and special conditions, e.g. moving machinery or chemical agents, the appropriate coping measurements can be taken. Moreover, the use of the incident detection system for a long period can provide statistics about the inherent dangers of the building. Experimental results seem very promising.
Keywords: collision, fall, incident detection, industrial environment, intrusion.

1 INTRODUCTION

The fourth industrial revolution, or else Industry 4.0, denotes a new manufacturing reality towards 'smart' factories. Such a factory could not omit the domain of occupational health and safety that until now mainly focused on the prevention of accidents by the use of area restrictions or appropriate equipment (e.g. helmets, gloves).

In this paper, following the vision of 'smart' industries, a real-time, multi-space incident detection system for indoor environments is introduced. The proposed system, which constitutes an extension of previous work described in Ref. [1], detects incidents such as falls, collisions and intrusions. The geometrical information of the BIM is utilized for the set-up and calibration of the sensors in the building permitting in this way to achieve multi-space detection. Once an incident is detected the appropriate coping mechanism is notified of the incident type and exact location on the building's architectural map so that it can immediately hasten to the point of the event and handle the situation. Combined with the knowledge of each area's particularities, which can be extracted from BIM, e.g. chemical agents, moving machinery, the appropriate coping measures can be taken. Moreover, statistics about the incidents occurrence in each area of the building can be easily made if the system is running for a long period leading to the appropriate readjustments – measures concerning the building. At this point, it should be mentioned that all utilized cameras are depth sensors respecting the personnel's legal and ethical rights.

The detection, recognition and modeling of the aforementioned incidents is accomplished through a Hidden Markov Model procedure that utilizes features that characterize each event, e.g. the vertical velocity of a falling human or the relative distance between two moving items that will collide.

To sum up, the main contributions of the proposed paper are:

1. The introduction of a real-time, multi-space incident detection system that produces alarms notifying for the type and exact location of the event on the building's architectural map so that the appropriate coping mechanism can be immediately triggered.

DOI: 10.2495/SAFE-V8-N2-266-275

This information combined with knowledge on each building's area particularities can lead to the appropriate decision of ways to handle the incident.
2. The introduction of a detection system that apart from falls and intrusions, which are incidents that have preoccupied a lot the research community, it supplementary detects collisions.
3. The incidents' modeling by a generic HMM procedure that utilizes characteristic features of each incident.
4. A tool that in long term can provide statistics about the inherent dangers of the building's areas so that measures are taken.

2 RELATED WORK

Fall detection is a topic with prolific literature. In general fall detection methods can be divided in two groups: 1) vision based algorithms, 2) techniques with non-vision sensors such as accelerometers [2]. The first category is mainly preferred since wearable equipment can be irritating for users. Moreover, depth sensors gain leverage in comparison to other vision sensors since they respect the legal and ethical rights of human privacy.

A usual criterion for fall detection methods is the distance from the centroid or top of a person to the floor. In particular, in [3] a predefined threshold to the floor and a specific time of immobility on the floor specify a fall while in Ref. [4] a fall is determined by a k-NN classifier that utilizes features such as head-floor distance, person area and shape's major length to width. Moreover, other approaches [5] utilize skeleton joints tracking so as to achieve fall detection. Furthermore, in Ref. [6] an ensemble of decision trees for the fall's confidence computation is proposed, while in Ref. [7] a Bayesian framework is used. A good survey on vision based approaches for fall detection can be found in [8].

Despite the fact that fall detection preoccupied plenty scientists, collision in indoor environments has not been explored, at least to our knowledge. There is work on collisions for traffic applications [9], nevertheless the environmental conditions and velocities in this situation are completely different.

3 INCIDENT DETECTION METHOD

The proposed incident detection method includes three steps. Firstly, the moving items in the monitoring area are detected and tracked. Subsequently, features that characterize the item's state, such as its velocity, area variance or distance from a region of interest are extracted. Finally, based on these features, an HMM method models, detects and recognizes the occurring incidents.

3.1 Detection and tracking of moving items

The first step of the method is detection and tracking of all the moving items in the monitoring area. In particular, it includes a robust, real-time, multi-camera, multi-space occupancy extraction system based on [10]. In this procedure, the geometrical building information extracted from the BIM is utilized for the appropriate set-up of the sensors and their calibration. Thus, all the utilized cameras' coordinates are transformed to a common global reference system located on the architectural map of the building enabling the monitoring of any area regardless its size. Furthermore, a virtual top-view camera based on the calibration data is utilized so that detection and tracking can be performed on the horizontal plane unaffected from partial occlusions. Moreover, a dual-band algorithm handles the dynamic changes of the environment by incorporating low objects, e.g. chairs, in a small time period, while retaining higher objects like humans for a longer period.

3.2 Feature extraction

The second step of the incident detection method is the extraction of features that characterize the state of the tracked items. These features depend on the type of incident under detection. Therefore, in the case of falls the selected features constitute the vertical velocity, area variance and height of the tracked worker while collision depends on the values of relative distance and velocity between two moving items. Contrary, intrusion is a straightforward procedure deriving from tracking therefore no further features are needed for its detection. In the following sections the fall and collision features are analyzed in detail.

3.2.1 Fall features

During a person's fall their vertical velocity and area measured from top view are augmenting drastically. In addition, their height at the end of the fall is relatively small. Therefore, these features are utilized to define a human at a falling state and are described below in detail.

1. Vertical velocity v: it constitutes the vertical velocity of the highest point of the tracked person's blob. Nevertheless, depth sensors' noise could affect the velocity's values leading to false results. Therefore, a six order, low pass FIR filter is applied on the height values while the velocity is averaged on a short constant window C_T. The velocity's formula is given below:

$$v = \frac{1}{C_T} \int_{t_c - C_T}^{t_c} h'(t)dt, \qquad (1)$$

   where $h'(t)$ is the tracked person's height derivative and t_c is the current time.

   Moreover, a big portion of the upper part of a human's blob could abruptly vanish due to partial occlusions resulting in false alarms. To deal with these cases the step signal created at the average velocity is detected and all the velocities that include it in their calculation are set to zero.
2. Area variance σ^2: it refers to the area of the tracked person from top view. It is obvious that as a person falls their area from top view drastically augments. To remain unaffected from the initial area before the fall the area variance is calculated in the same time window C_T as vertical velocity assuring relative robustness to noise. Its formula is described by the equation:

$$\sigma^2 = \frac{1}{C_T} \int_{t_c - C_T}^{t_c} \left(A(t) - \int_{t_c - C_T}^{t_c} \frac{A(t')}{C_T} dt' \right) dt, \qquad (2)$$

   where $A(t)$ represents the area of the tracked person from top view.
3. Height h: it is the highest point of the tracked person's blob. The value of the final height of the person after the fall facilitates the avoidance of false alarms.

3.2.2 Collision features

Contrary to fall, collision needs more than one person or moving item in order to occur. Therefore, the features that define it are relative features between two participants. In particular the relative distance and velocity between two objects are utilized.

1. Relative distance d_r: it refers to the distance of two participants on the horizontal plane. As it is obvious the magnitude of the distance highly relates to the probability of collision.
2. Relative velocity v_r: it represents the speed the two participants are approaching or diverging, affecting the collision probability. It is given by the following formula:

$$v_r = \frac{d(dr)}{dt}. \tag{3}$$

3.3 Incident recognition

A three state HMM is utilized for modeling and recognizing the incidents based on the aforementioned features. In both cases the first state (S1) defines a stable condition where the incident under detection has not occurred yet. The second state (S2) includes the actual action of the incident while the third state (S3) signifies the termination and confirmation of the event. For example, a walking worker would be in state S1. Once he started falling he would enter S2 while at the end of the fall, once he reached the floor, he would be at state S3 unless if in the meantime he regained his balanced and returned to state S1. In the same spirit, two moving objects at a big distance from each other would be in state S1. Once they started approaching each other in a close distance they would transfer to state S2 and if they finally collided they would enter state S3.

The transition probabilities of the HMM are encapsulated in the following formula:

$$P = \begin{bmatrix} (1-E) & E & 0 \\ (1-E)(1-u) & E & (1-E)u \\ r & 0 & (1-r) \end{bmatrix}, \tag{4}$$

where E, u and r are functions that depend on the event. Nevertheless, in both cases of fall and collision, function E describes the actual incident as it occurs; function u refers to the termination and verification of the event while function r depends on the time needed to transfer to another state after the occurrence of the event. The functions are analytically described for each incident separately (depending on the event each function is represented with the corresponding subscript).

3.3.1 Fall

As it is already mentioned, the characteristic features of human fall are the vertical velocity v, the area variance σ^2 and the final height at the end of the fall h. Therefore the functions of the transition probabilities depend exclusively on these characteristics and can be described by the following formulas:

$$E_f = \frac{1}{q + e^{v\sigma^2 + \mathrm{T}_f}}, \tag{5}$$

$$u_f = \begin{cases} 1, & H_{T_f} - h \geq 0 \\ 0, & H_{T_f} - h < 0 \end{cases}, \tag{6}$$

$$r_f = 1, \tag{7}$$

where T_f and H_{T_f} are constants.

The eqn (5) is a sigmoid function that represents the probability E_f since for cases of high vertical velocity and area variance it reaches high values close 1. Moreover, eqn (6) is a step function that defines the detection of a fall since a person cannot be considered fallen if their height is above a predefined threshold H_{T_f}. Finally, eqn (7) signifies that once the fall has ended the HMM notifies of the incident and transfers to a stable condition.

3.3.2 Collision

The characteristic features of collision are the relative distance and velocity of the two participants to the event. The functions that describe this incident are the following:

$$E_c = e^{-\sigma_{c_d}(d_r - T_d)} \frac{1}{1+e^{\sigma_{c_v} v_r}}, \tag{8}$$

$$u_c = \begin{cases} 1, & D_T - d_r \geq 0 \\ 0, & D_T - d_r < 0 \end{cases}, \tag{9}$$

$$r_c = 1 - u_c, \tag{10}$$

where σ_{c_d} and σ_{c_v} define the slope of the two functions comprising the eqn (10), D_T and T_d are constants.

Equation eqn (8) comprises two functions that refer to the relative distance and the relative velocity respectively. The exponential function represents the inversely analogous relationship between the probability of collision and the distance between the two participants while the sigmoid function depicts the fact that highly negative velocities decrease the probability of collision while highly positive increase it. The step function in eqn (9) determines the distance that signifies a collision while eqn (10) declares that a collision is considered to have ended only when the distance between the two participants is above a predefined threshold.

4 EXPERIMENTS

4.1 Datasets

Two datasets from two different work environments at Centre of Research and Technology Hellas (CERTH) were acquired for the evaluation of the proposed method. The datasets comprised various incidents including falls and collisions performed by 28 different participants. Nevertheless, other events that resemble the incidents under detection, such as bending to pick up something or tie one's shoes, two persons walking to the same direction without colliding etc, are also reproduced in order to evaluate the robustness of the algorithm in incidents that affect the features that determine the HMM transitions. Screenshots from the datasets are depicted in Fig. 1.

The dataset consists of 140 falls, 60 collisions, 50 cases of people bending to tie their shoes, 100 cases of people bending to pick up an item (e.g. various sizes of boxes, chairs) and 40 cases of pairs of people walking by each other without colliding. In general, the participants were allowed to conduct the incidents in ways that felt natural to them resulting in a variety of data. Therefore, there were people falling forwards, backwards or sideways, while standing, as if they were fainting, or while walking, as if they were stumbling. Moreover, the participants simulating collision diverged and approached in various directions and velocities. Since the experiments were conducted in work environments there were many members of the personnel walking or standing in the monitored area but although they did not produce false alarms they were not taken into account for the method's evaluation.

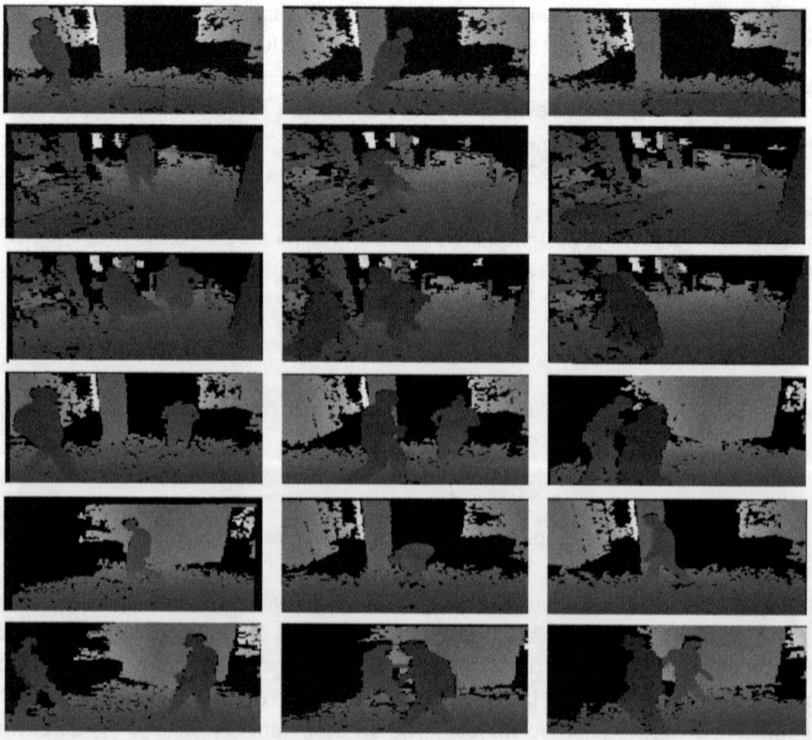

Figure 1: Incidents included in the dataset. The first two rows depict fall events, the third and forth include collision incidents while the fifth row shows an event similar to fall, i.e. bending, and the last row an event that resembles collision, meaning two persons passing by each other in a close distance.

4.2 Feature processing

In this section the features that characterize each incident and determine the HMM transition probabilities are further analysed with some examples so that their influence and importance on incident detection is better displayed. Thus, in Fig. 2 the vertical velocity, area variance and height during a fall are depicted, while Fig. 3 illustrates the relative distance and velocity during a collision. In both cases the transition probability function E (eqns (5) and (8)) that represents the actual fall is depicted.

In particular, Fig. 2 represents the height of a worker during a fall incident. As it can be observed, at the beginning the worker is in standing position, then, falls and immediately starts rising. In correspondence to the height diagram, vertical velocity (whose direction is defined upwards) starts from zero values, drastically decreases during the fall and, finally, fluctuates near zero. Similarly, the area variance highly augments during the fall. Based on these values, the probability function reaches high values during the fall while retains low values in the rest time period. The sequence of the HMM states according to the probability of values is depicted with colours on the graphs.

In Fig. 3 graphs of collision features are illustrated. As the moving objects approach their relative distance diminishes and their relative velocity takes negative values. Once the relative distance reaches and remains below a threshold the relative velocity oscillates due to collision.

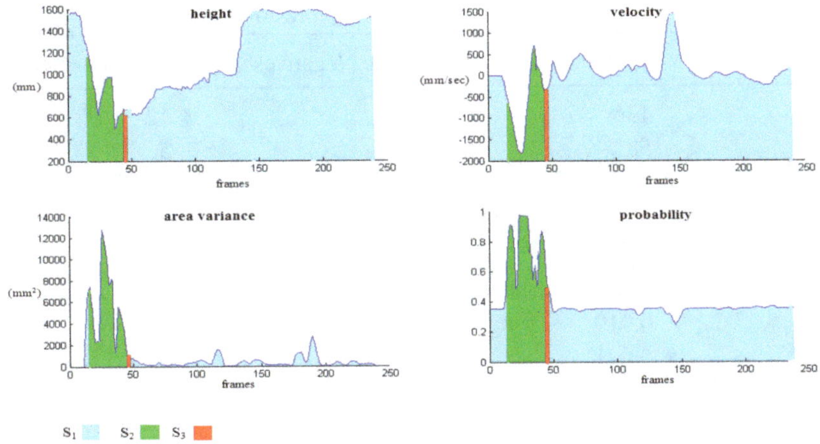

Figure 2: Fall features and probability E_f. The sequence of the HMM states according to the probability values is depicted with colours.

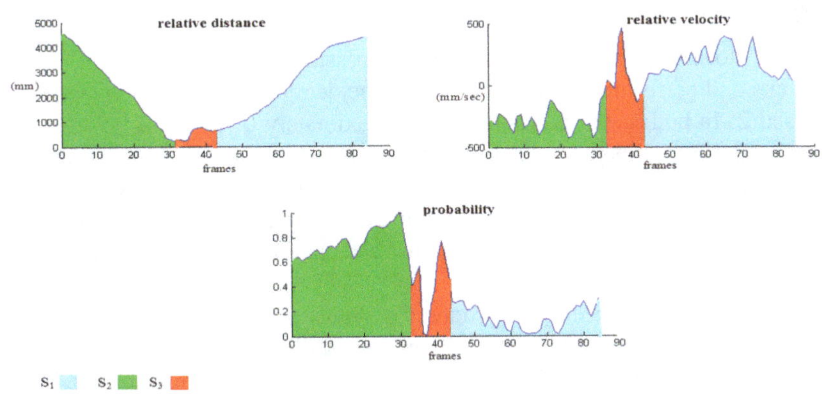

Figure 3: Collision features and probability E_c. The sequence of the HMM states according to the probability value is depicted with colours.

At this point the HMM state transitions from S2 to S3, thus probability E_c does not affect the method. When the collision terminates the state transitions from S3 to S1 and probability E_c remains below 0.5. It is obvious that although the relative distance's values are similar before and after the collision probability E_c diminishes drastically due to relative velocity.

4.3 Experimental results

The proposed method was tested in more than 600000 frames that included apart from the incidents of interest, i.e. falls and collisions, events with similar properties to test the algorithm's robustness. The extraction of the results is not frame-based but depends on the number of times each incident occurred.

Fall incidents were tested in 140 cases. In 136 cases the falls were detected correctly, 4 were missed due to step signal just before the fall while 12 events similar to fall produced

Table 1: Incident detection results.

	True Positives	True Negatives	False Positives
Fall	136	238	12
Collision	58	325	5

Table 2: Incident detection performance.

	Precision	Recall	F1-score
Fall	0.919	0.971	0.945
Collision	0.921	0.967	0.947

false alarms. Nevertheless, it should de mentioned that 6 of the false positive cases referred to bending to pick up a mattress used for falls. However, incidents of picking up such a voluminous object do not occur often in industrial environment or are limited to a certain area.

In the incident of collision, 58 out of 60 cases were detected correctly, 2 were missed while 5 false alarms were produced from similar events (3 of them referred to persons meeting and standing very close to each other and 2 to persons passing by each other).

The results and performance of the presented incident detection method are analyzed in Tables 1 and 2. In both cases precision was approximately 0.92, recall 0.97 and F1-score 0.95, which can be considered very promising results. As it is already mentioned the proposed incident detection system can also support detection of intrusion to restricted areas. Nevertheless, since the detection of the particular incident depends completely on the accuracy of the tracking algorithm, which is explored and analyzed in Ref. [11], no experimental results are provided in this paper.

All possible alarms produced by the proposed system are depicted in Fig. 4. The location of the incident on the architectural map of the building and the input from two depth cameras monitoring the scene are illustrated. Each incident is depicted with different color. In Fig. 4a–c fall, collision and intrusion to a restricted area (marked with a gray rectangle) are depicted. In Fig. 4d possible collision between two moving objects is depicted. This alarm could serve as a proactive notification in order to avoid accidents and corresponds to state S2 of the HMM.

5 CONCLUSIONS

This paper presents a real-time, multi-space incident detection system for indoor environments. The system focuses on industrial environments while falls, collisions and intrusions to restricted areas constitute the incidents under detection. Once an incident is detected, an alarm is produced notifying for its type and exact location so that a coping mechanism can be immediately triggered. The importance of this prompt notification increases when the incident occurs in remote or hazardous areas with chemical agents, hard climatic conditions or running machinery. Therefore, the combined knowledge of the incident area's conditions and its type provide leverage for the building's personnel safety and health since the appropriate measures can be immediately taken. Furthermore, in long term, the system could provide statistics for the occurring incidents so that adjustments are made for better incident prevention. The importance of BIM in this application is high since it constitutes an essential component for the set up of the system while it provides information about the location of the

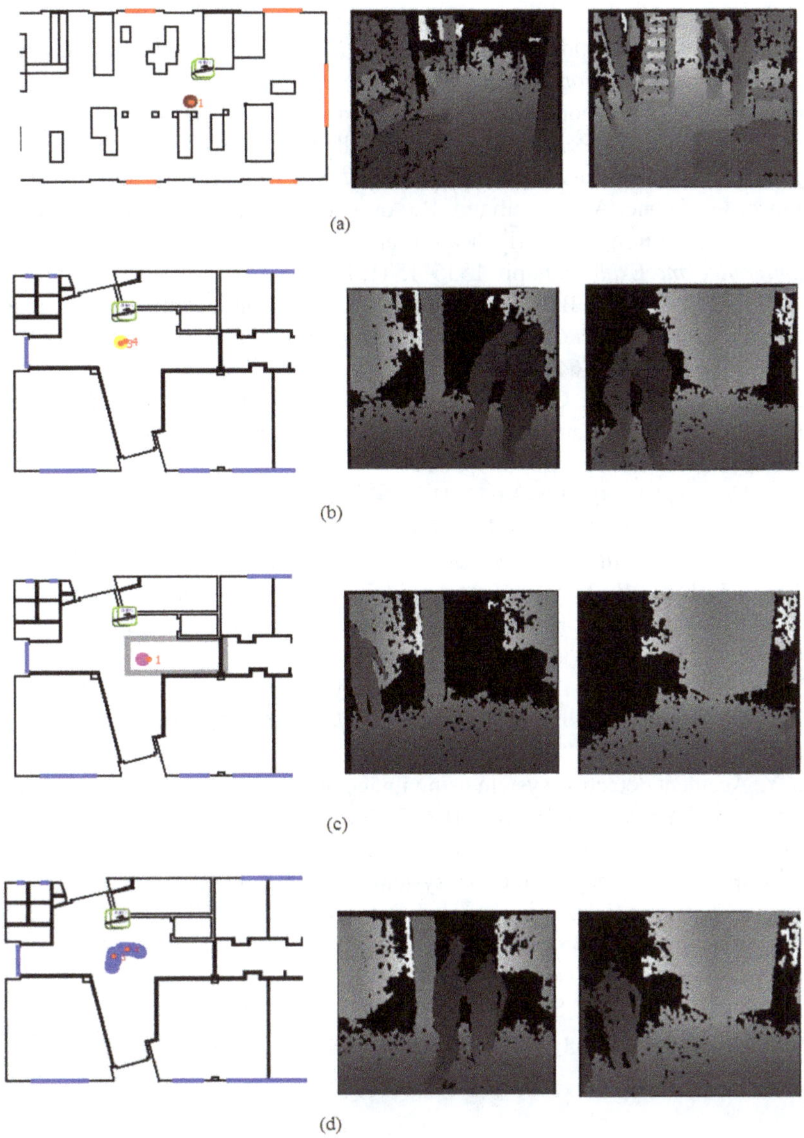

Figure 4: Incident detection and indication of their location at the building's architectural map: (a) fall, (b) collision, (c) intrusion to restricted area, (d) indication of possible collision.

incidents facilitating the immediate and appropriate coping actions according to the area's particularities.

Moreover, the proposed method for incident detection is based on a generic HMM procedure that utilizes features that characterize the incidents under detection, e.g. vertical velocity, relative distance between two moving items etc. The experimental results are very promising and conform to the vision of a 'smart' and safer industrial environment.

REFERENCES

[1] Triantafyllou, D., Krinidis, S., Ioannidis, D., Metaxa, I.N., Ziazios, C. & Tzovaras, D., *A real-time fall detection system for maintenance activities in indoor environments*, 3rd IFAC AMEST Workshop, 2016.

[2] Wu, G. & Xue, S., Portable pre-impact fall detector with wearable sensors. *IEEE Transactions on Neural Systems and Rehabiliation Engineering*, pp. 178–183, 2008.

[3] Diraco, G., Leone, A. & Siciliano, P., An active vision system for fall detection and posture recognition in elderly health care. *Design, Automation and Test in Europe Conference and Exhibition*, pp. 1536–1541, 2010.

[4] Kepski, M. & Kwolek, B., Fall detection using ceiling-mounted 3d depth camera. *International Conference on Computer Vision, Imaging and Computer Graphics Theory and Applications*, pp. 640–647, 2014.

[5] Zhang, C., Tian, Y. & Capezuti, E., Privacy preserving automatic fall detection for elderly using RGBD cameras. *International Conference on Computer Helping People with Special Needs*, pp. 625–633, 2012.
https://doi.org/10.1007/978-3-642-31522-0_95

[6] Stone, E. & Skubic, M., Fall detection in homes of older adults using the Microsoft kinect. *IEEE Journal of Biomedical and Health Informatics*, pp. 1590–1594, 2014.

[7] Zhang, Z., Liu., W., Metsis, V. & Anthitsos, V., A viewpoint- independent statistical method for fall detection. *IEEE Conference on Pattern Recognition*, pp. 3626–3630, 2012.

[8] Zhang, Z., Conly, C. & Anthitsos, V., A survey on vision based fall detection. *International Conference on Pervasive Technologies Related to Assistive Environments*, 2015.
https://doi.org/10.1145/2769493.2769540

[9] Ki, Y., Accident detection system using image processing and MDR. *Computer Science and Network Security*, **3**(7), pp. 640–647, 2007.

[10] Krinidis, S., Stavropoulos, G., Ioannidis, D. & Tzovaras, D., A robust and real time multi space occupancy extraction system exploiting privacy preserving sensors. In *Communications, Control and Signal Processing (ISCCSP), 2014 6th International Symposium*, pp. 542–545, 2014.
https://doi.org/10.1109/isccsp.2014.6877932

[11] Ioannidis, D., Krinidis, S., Drosou, A., Tzovaras, D. & Likothanassis, S., Occupant-aware indoor monitoring for enhanced building analysis. *SimAUD15*, pp. 85–92, 2015.

IMPROVING THE DYNAMIC PERFORMANCE OF MULTI- STOREY BUILDINGS VIA PROTECTIVE GLAZING CURTAIN WALLS

CHIARA BEDON & CLAUDIO AMADIO
University of Trieste, Department of Engineering and Architecture, Trieste, Italy.

ABSTRACT
Glazing facades are widely used in building structures, due to a series of aesthetic, thermal, lightening aspects. From a structural point of view, under the action of exceptional loads as impacts, explosions, seismic events or hazards in general, the glazing envelopes often represent the critical component for multi-storey buildings, due to the typically brittle behaviour and limited tensile resistance of the glass panes, hence requiring specific, fail-safe design concepts. In this paper, the feasibility and potential of special mechanical connectors interposed at the interface between a given multi-storey primary building structure and the glazing facade are investigated via accurate finite-element numerical models, under various impact scenarios. As shown, the final result is an assembled structural system in which the facade can work as passive control system for the primary structure, in the form of a distributed tuned-mass damper (TMD).
Keywords: blast scenarios and hazards, dissipative devices, Finite-Element numerical models, glazing curtain walls, multi-storey buildings, passive structural control, tuned-mass dampers (TMDs).

1 INTRODUCTION AND MOTIVATION

Glazing facades are widely used in building structures, due to a series of aesthetic, thermal, lightening aspects. In most of the cases, wide transparent surfaces are created in commercial, residential as well as strategic buildings, including airports, museums, offices, etc.

From a structural point of view, however, under the action of exceptional loads as impacts, explosions, seismic events or hazards in general, the glazing envelopes often represent a critical component for multi-storey buildings, due to the typically brittle behaviour and limited tensile resistance of glass panes, hence requiring specific, *fail-safe* design concepts [1]. The appropriate estimation of the vulnerability of glazing systems, as well as the prediction of their actual dynamic behaviour under exceptional loads (including the interaction between a given envelope and the substructure/primary building), or the implementation and development of advanced retrofitting and enhancing techniques, consequently, are currently open topics still requiring extended studies.

In this paper, taking advantage of major outcomes of an ongoing research investigation, the feasibility and potential of special mechanical connectors interposed at the interface between a given multi-storey primary building structure and the enclosing glazing facade are investigated via accurate finite-element (FE) numerical models, under various impact scenarios. Careful consideration is paid, in particular, for the mitigation of effects due to seismic loads and blast events. As shown, when properly designed, the proposed special connectors can markedly improve the overall dynamic performance of a given traditional structure, both in terms of global building response and local performance (i.e. at the facade components level). The final result, consequently, potentially consists in an assembled structural system in which the glazing facade can work as a passive control system for the primary structure, in the form of a distributed tuned-mass damper (TMD).

DOI: 10.2495/SAFE-V8-N2-276-286

2 DESIGN REGULATIONS FOR GLASS FACADES UNDER EXCEPTIONAL LOADS

The design of structural glass elements and facades in general under exceptional loads is strictly dependent on the actual performance (i.e. stiffness, resistance, etc.) of single structural components, as well as their reciprocal interaction under the assigned loading combination. As such, careful consideration should be paid also for anchoring systems, supports, framing members, etc. Glass fragments represent in fact a critical issue for people, hence cracking of panes should be generally prevented (i.e. Fig. 1(a) and (b)). On the other hand, an optimal combination of safety levels, global and local structural performances, costs should be also achieved.

As far as the given glass system to verify can be considered as a *secondary* component, compared to the primary structure, partial damage is generally accepted by available design standards. This is not the case of structural glass assemblies of *primary* importance within a given structural system. There, the glass elements or facade components should in fact able to properly resist to the incoming impulse, as well as to accommodate the overall deformations of the building as a whole, including both out-of-plane and in-plane displacements. In the latter case, it is hence clear that special mechanical connectors and fasteners are mandatory to satisfy design regulations and avoid severe damage.

General European standards for the seismic design of buildings can be applied to glass curtain walls (see, for example, Ref. [2]). There, secondary components only are considered, however, and no specific design regulations are available to account for the importance or typology the curtain wall belongs, as well as for detailing, anchoring systems, etc. More detailed provisions are included in FEMA 450 [3], even for secondary non-structural cladding systems only, where drift limit values to satisfy in order to avoid glass fallout are specified for glazed curtain walls, storefronts and partitions.

Regarding the design of blast resistant windows and facades in general, EU as well as US standards are available for testing laminated glass windows [4, 5], and hence to define the corresponding hazard level. The latter one, due to glass intrinsic brittleness, is typically measured by the fragments that are found after the experiment behind the glass pane (see Fig. 1(c)).

3 EXPLORING THE DESIGN CONCEPT OF DISTRIBUTED TMD FACADES

3.1 Theoretical background

The design concept of building structures equipped by vibration control devices and dissipative glazing curtain walls (GCWs) takes inspiration from the original TMD concept, as

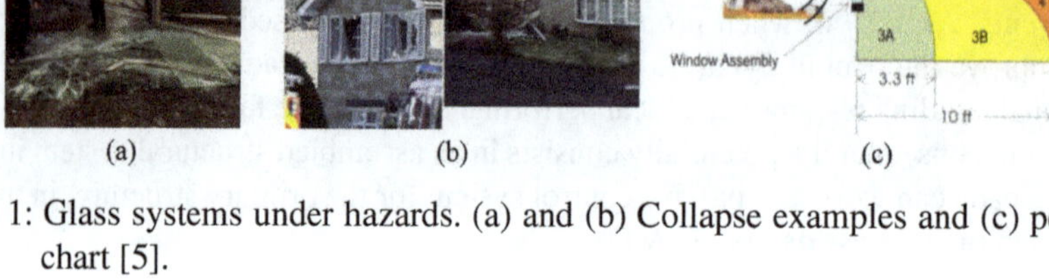

(a) (b) (c)

Figure 1: Glass systems under hazards. (a) and (b) Collapse examples and (c) performance chart [5].

well as from the use in traditional facades of special fasteners able to act as passive impact absorbers (Fig. 2).

The use of vibration control systems in facades for the enhancement of tall buildings performance under wind and seismic loads has been first theoretically explored in Ref. [6], with attention for double skin (DS) facades. There, DS facades with special connectors proved to reduce the seismic effects on tall buildings (up to -35%). Further applications of special connectors have been proposed for cable supported facades under explosions [7, 8], while special viscoelastic (VE) or ADAS brackets for GCWs have been numerically investigated in Ref. [9], giving evidence of their potential. In this paper, the design concept of GCWs acting as passive absorbers for buildings under several hazard typologies is further investigated, via computationally efficient but accurate FE models.

The dynamic performance of a building with GCW and vibration control devices can be generally described as shown in Fig. 2 (detail box). The primary structure (total mass M_{struct}, stiffness K_{struct}, damping c_{struct}) interacts with the GCW – namely representative of additional mass $M_{glass,tot}$ but limited stiffness, for the primary structure – via special connectors replacing the usually rigid brackets ($K_{d,tot}$ and $c_{d,tot}$ their total stiffness and damping). The dynamic behaviour of such a SDOF-TMD under a given design load, in particular, is described by [6]:

$$M_{struct}\ddot{u}(t)+C_{struct}\dot{u}(t)+K_{struct}u(t)=P(t)+C_{d,tot}\dot{u}_d(t)+K_{d,tot}u_d(t) \tag{1}$$

$$M_{glass,tot}\ddot{u}_d(t)+C_{d,tot}(\dot{u}_d(t)-\dot{u}(t))+K_{d,tot}(u_d(t)-u(t))=p(t) \tag{2}$$

where $u(t)$ is the deflection in time of the primary structure; $u_d(t)$ denotes the GCW relative displacement; $P(t)$ and $p(t)$ represent the force acting on the building mass or on the TMD mass, respectively.

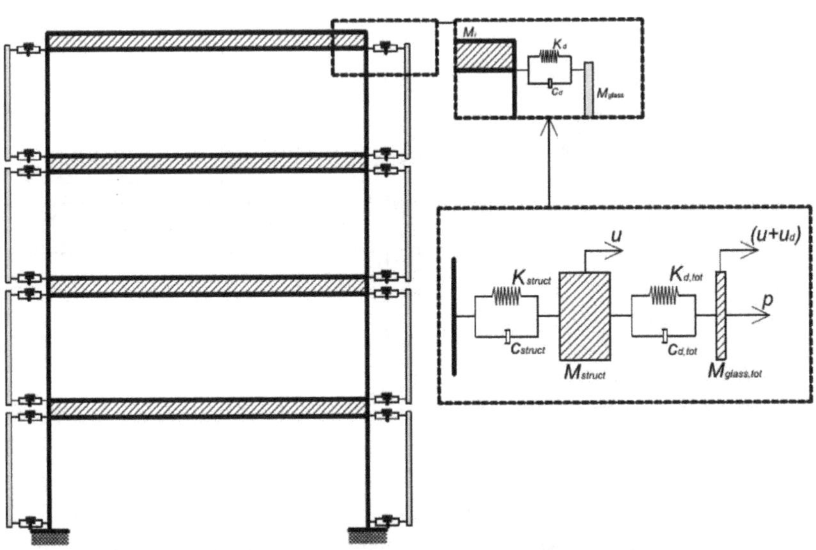

Figure 2: Reference mechanical model for the analysis of a building with dissipative GCW.

3.2 Design considerations and prototype

Assuming that the single GCW modular unit is connected to the adjacent primary frame by means of four fasteners (i.e. one device at each panel corner), the preliminary estimation of $K_{d,tot}$ and $c_{d,tot}$ can be carried out based on Fig. 3. There, M_{glass} is the total mass of a single GCW panel (plus the supporting frame), while K_d and c_d denote the stiffness and damping terms for a single device.

The typical control system can be composed by a VE solid damper, consisting of metal plates and a middle rubber layer (h_d its thickness and A_d the surface area). The device base is directly attached to the structural backup (e.g. concrete slab) by means of anchoring bolts, whereas a further sliding steel plate supports the GCW framing members, enabling possible crushing and rotations of the VE layer. To this aim, at the interface between the VE compound and the sliding steel bracket, additional gaskets are interposed, so to avoid direct contact and infinitely rigid restraints for the sliding VE layer.

The prototype of Fig. 3 takes inspiration from classical viscous dampers in use for the seismic mitigation of tall buildings, see Ref. [10]. For design purposes, key parameters under well-defined loading conditions (e.g. operating frequency ω and temperature T) are the damping ratio c_d deriving from the rubber compound and the corresponding stiffness K_d, where [10]:

$$c_d = c_d(\omega) = \frac{K^{"}}{\omega} = \eta \frac{K^{'}}{\omega}, \tag{3}$$

with η the loss factor of the VE layer and

$$K^{'} = \frac{G^{'}(\omega) A_d}{h_d}, \; K^{"} = \frac{G^{"}(\omega) A_d}{h_d} \qquad (4), (5)$$

the storage and loss stiffnesses, where $G'(\omega)$ and $G''(\omega)$ denote the corresponding shear moduli.

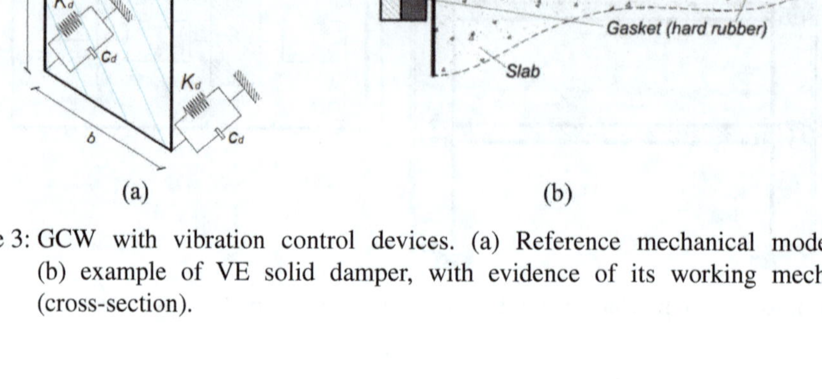

(a) (b)

Figure 3: GCW with vibration control devices. (a) Reference mechanical model and (b) example of VE solid damper, with evidence of its working mechanism (cross-section).

Assuming a high damping rubber compound (ξ = 20%), an average value G' = 0.35MPa was taken into account [11, 12]. In terms of gaskets, conversely, an hard compound was considered (G' = 1.2MPa), with damping contributions fully neglected. Given the fundamental period T_1 of the primary structure, a preliminary estimation of K_d can be rationally carried out as:

$$K_d = \frac{1}{4} K_{d,glass} = \frac{M_{glass} \pi^2}{T_{1,glass}^2} \tag{6}$$

with $T_{1,glass} = T_1$, which is by equaling the vibration period of the GCW modular unit and the building one. The corresponding damping term is then given by [13]:

$$c_d = \frac{1}{4} c_{d,glass} = \frac{1}{4}\left(\xi \cdot c_{cr}\right) = \frac{1}{4}\left(\xi \cdot 2M_{glass}\omega_{1,glass}\right), \tag{7}$$

with $w_{1,glass}$ being assumed to coincide with the damper operating frequency, ξ the VE damping coefficient and c_{cr} its critical value.

Finally, see Fig. 3, possible failure mechanisms in the VE layer (i.e. tearing) under a given design load should be properly prevented, by limiting its maximum shear deformations:

$$s_d \leq s_{d,\max} = \min\left(2h_d;\ 30\right) \text{ in [mm].} \tag{8}$$

4 FINITE-ELEMENT NUMERICAL INVESTIGATION

4.1 FE modelling approach and solving strategy

In this paper, the feasibility and potential of the proposed VE connectors is assessed via FE numerical simulations. Compared to earlier research efforts, careful consideration is given to a given GCW module under the effects of various types of impact loads, including blast events, as well as seismic loads. To this aim, simulations are carried out at the level of a full 3D building, as well as in terms of single GCW components, so to explore global and local effects.

Beside the source and features of the input load, in any case, the same general FE modelling approach was taken into account and the typical FE model was assembled and calibrated so to represent a traditional GCW modular unit.

In order to maximize the computational efficiency of FE models but preserve their accuracy, the GCW unit was numerically described in the form of *shell* elements, *beam* elements and mechanical joints, see Fig. 4. A total number of 6000 elements was used to represent the glass pane and the supporting frame, with 35000 DOFs. For the glass panel, 4-node and 3-node monolithic shell elements (S43, S3R) were defined, with a total thickness of 24 mm [9]. Based on a free meshing technique, the size of shell elements was refined in the central region of the pane (i.e. where cracks are expected) and optimized towards the edges, see Fig. 4(b), with average length comprised between 0.004 m and 0.1 m. The metal frame was described via 1D *beam* elements (B31 type), with box cross-sectional shape well representative of a typical framing system for GCWs [9]. In the latter case, the mesh size was set to 0.1 m, so to have an optimal correspondence between mesh nodes lying on the glass panel edges and along the frame members.

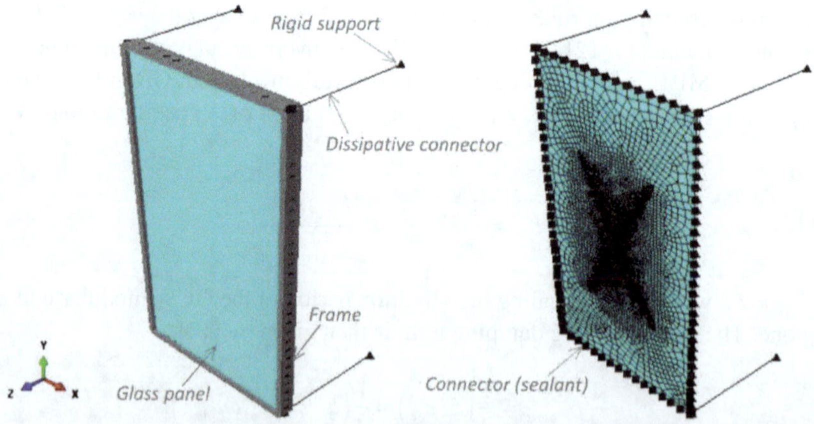

Figure 4: FE model assembly and features. (a) Extruded view and (b) overview of mesh patterns and mechanical connectors (ABAQUS [14]).

A key role was then assigned to *joints*, being representative of the actual mechanical interaction between the glass panel and the frame (where sealant joints and gaskets are usually interposed between them), as well as of the structural behaviour of proposed VE fasteners.

In the first case, a combined *'cartesian + rotation'* joint was assigned to all the glass panel/frame nodes. Aiming to take into account metal gaskets and frame detailing, relative rotations were restrained between the involved nodes, while a brittle elastic behaviour was defined to represent the in-plane stiffness of sealant joints [9].

To reproduce the mechanical response of VE connectors, nonlinear joints (*'cylindrical'* type of ABAQUS library [14]) were indeed linked to the frame corner nodes, see Fig. 4. Stiffness and damping properties for the VE compound were estimated by taking into account eqns (6) and (7). The presence of hard rubber gaskets was also properly considered.

In terms of materials, the *brittle cracking* damage model was used for glass, so to take into account possible tensile cracking of the panel. Input parameters were calibrated in accordance with Ref. [9]; with 85 MPa, the reference tensile resistance of annealed float glass under impulsive loads and E_g = 70 GPa the modulus of elasticity. An elasto-plastic constitutive law was indeed then defined for the metal frame, with E_a = 70 GPa, $s_{y,a}$ = 200 MPa, $s_{u,a}$ = 280 MPa the nominal reference values for Young modulus and yielding/ultimate stresses [9].

4.2 Seismic loads

The potential of GCWs with VE fasteners in presence of seismic loads was first considered, by taking into account a full 3D system, being representative of a steel framed, 3-storey building, see Fig. 5(a).

To this aim, the typical GCW unit shown in Fig. 4 was reproduced in series, to describe a full facade (see details of Fig. 5), and properly restrained to the steel frame. Nonlinear dynamic simulations were then carried out by imposing a set of seven seismic records at the building base. All the records, consisting of two-component acceleration data, were derived to be consistent with an EC8 [2] acceleration spectrum associated to type A soil (rock soil), T1 topographic category, 0.35 g peak ground acceleration and 50 years of nominal life (ultimate limit state).

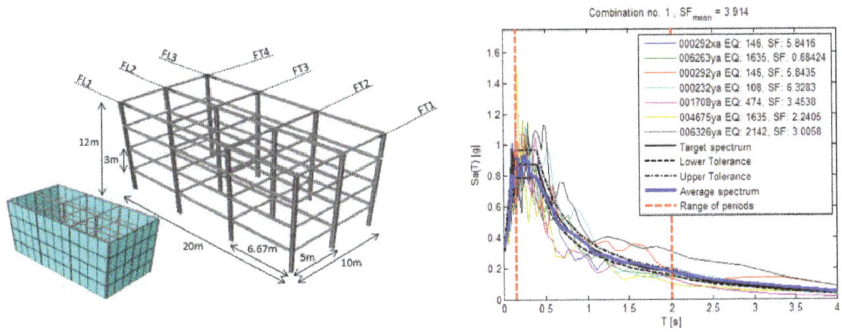

(a) (b)

Figure 5: (a) Full 3D steel building (ABAQUS [14]) and (b) set of seismic acceleration spectra.

Following section 3, K_d was first estimated as a function of the building period $T_1 = 0.326$s, as numerically calculated. A non-dimensional magnifying coefficient R_K was then also defined, so that the actual stiffness of a single VE device could be calculated – through the FE parametric study – as (with $1 \leq R_K \leq 50$, in this study):

$$K_d = \left(\frac{M_{glass} \pi^2}{T_1^2} \right) \cdot R_K \tag{9}$$

Figure 6 summarizes the typical effect of the proposed VE devices, for the examined building under seismic events. As far as K_d modifies, in accordance with eqn (9), a certain flexibility and damping capacity is progressively introduced in the traditional frame with fully rigid brackets (i.e. $R_K \geq 50$).

In terms of global performance of the building, beside the assigned R_K value, the fundamental vibration shape was found to slightly modify, see Fig. 6(a). At the same time, limited variations were observed in terms of fundamental period T_1. However, due to the additional flexibility and damping terms deriving from VE fasteners, major benefits were indeed noticed in terms of mitigation of seismic effects on the main building as well as on glass panels.

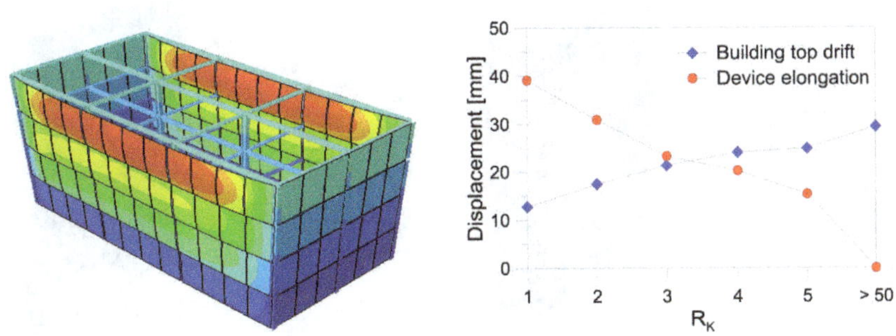

(a) (b)

Figure 6: Seismic performance a 3D building with GCWs and VE devices (ABAQUS [14]). (a) Fundamental vibration shape and (b) top drift/ VE device elongation, by varying K_d.

In Fig. 6(b), for example, maximum top drift amplitudes measured for the full 3D frame under the assigned set of seismic accelerations are proposed, for some selected FE models, and compared with the corresponding elongation of VE devices. As shown, a $R_K \geq 50$ coefficient was found to coincide with almost fully rigid brackets for the GCW, hence the benefits of VE fasteners expected to vanish. As far as R_K decreases, the combination of additional flexibility and damping capacity due to VE connectors manifests in a marked reduction of the measured building top drift, with progressively increasing sliding demand for the VE devices. In this regard, based also on design concepts summarized in section 3, parametric FE studies partly discussed in this paper proved to offer – for the examined case study – maximum benefits of VE devices for R_K values equal to ≈3–5.

In terms of dynamic response of the single GCW module and VE effects on the unit components, further comparative FE results are proposed in Fig. 7, with evidence of tensile stresses amount and distribution in glass under the assigned set of ground accelerations. Due to combined flexibility and damping properties of VE devices (when properly designed), part of the incoming seismic input energy is preliminary dissipated. As a result, the single GCW unit as well as the bare frame supporting the facade are subjected to a reduced impulse, compared to the same 3D system with fully rigid brackets.

In this sense, maximum peaks of stresses in glass are first highly mitigated by VE devices (see Fig. 7(a)). At the same time, the additional flexibility of GCW supports also beneficially affects the overall distribution of maximum stresses in the glass panels. As shown in Fig. 7(b) and (c), local peaks of stresses close to the glass supports and edges of glass can be generally avoided, thanks to VE fasteners, with obvious benefits for the overall dynamic performance of the full structural system.

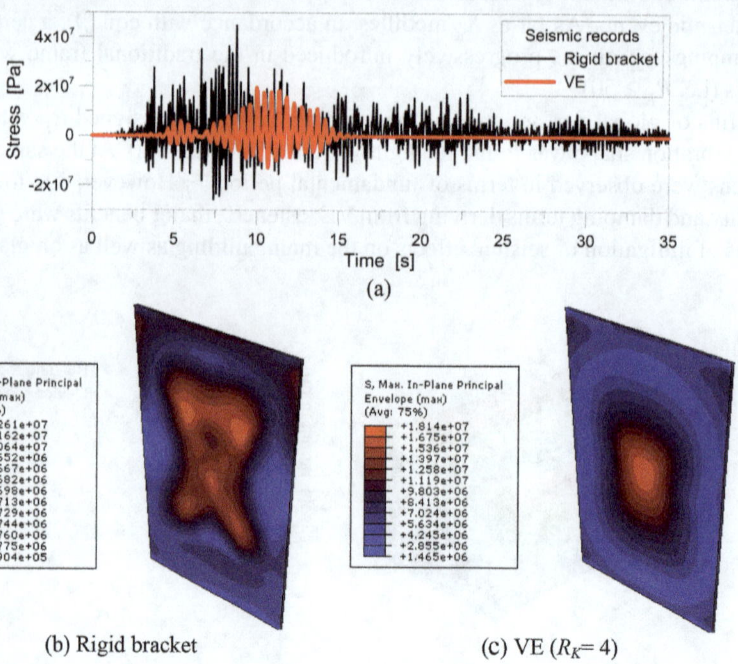

Figure 7: Tensile stresses in glass for the GCW unit under seismic events (ABAQUS [14]). (a) Maximum envelope (panel center) and (b)–(c) contour plot (t = 10 s, with legend in Pa).

4.3 Blast loads

By taking into account the same GCW unit of Fig. 4, its dynamic response under blast pressures was then also investigated, in order to further assess the potential of proposed VE devices.

Based on section 4.2, the $R_K = 4$ configuration was first taken into account, since representative of a rather optimal balance of hazard mitigation and VE device performances for the case study building under seismic events.

In particular, several input blast waves were taken into account by considering a certain amount of equivalent TNT charge and a given stand-off distance of 30 m from the facade.

Both the positive and negative phases were considered, in accordance with [9]. The so defined blast pressure waves were then applied on the glass surface in the form of uniform, time-varying pressures. In Fig. 8, major outcomes of selected FE simulations are proposed.

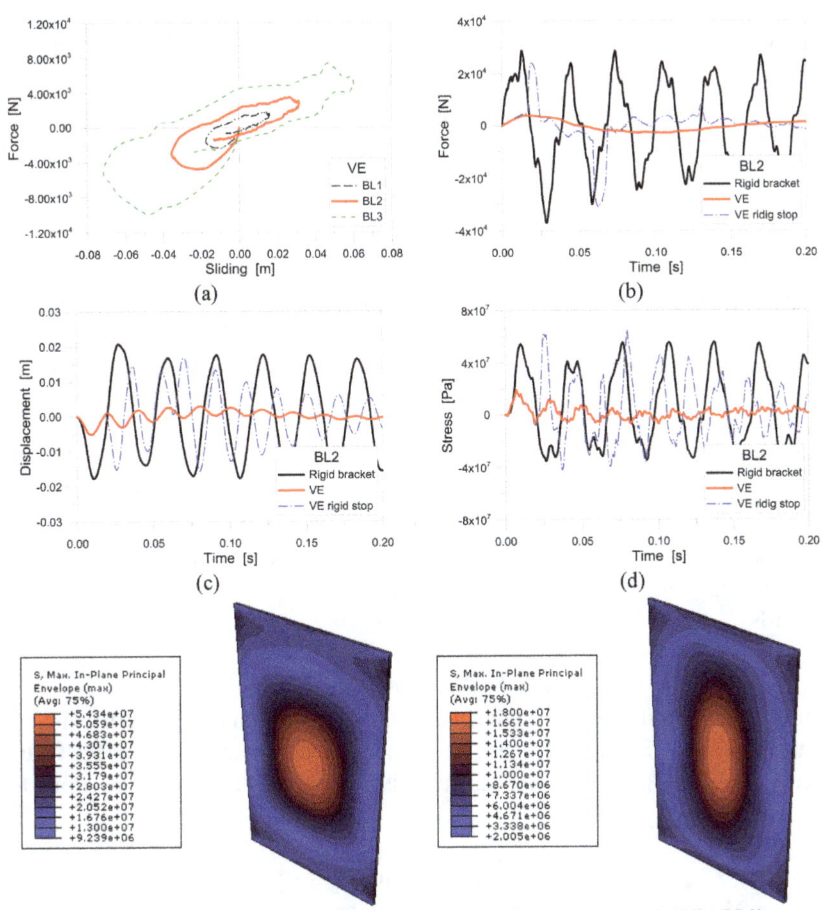

(e) Rigid bracket (t= 0.01s, BL2) (f) VE (R_K= 4, t= 0.007s, BL2)

Figure 8: Response of the GCW unit under blast loads (ABAQUS [14]), with FE results proposed for the $R_K = 4$ configuration. (a) VE connector behaviour; (b) reaction forces at the supports, as a function of time; (c) relative displacements and (d) maximum stresses at the center of glass. (e)–(f) Blue-to-red contour plot of tensile stresses in glass (legend in Pa).

There, the loading configurations denoted as BL1, BL2 and BL3 correspond, respectively, to a charge of 12.5 Kg, 25 Kg and 50 Kg of explosive, being representative of blast loads B (low level), C (medium) and D (high), as defined by design standards [5].

As shown in Fig. 8(a), rather stable behaviour was generally observed for the VE devices, even under impulsive blast waves typically characterized by abrupt release of energy and limited duration, in the order of thousand of milliseconds. Under the BL2 loading configuration, in particular, it can be seen in Fig. 8(b) to (d) that the actual effects of the incoming blast pressure are again markedly reduced, compared to the traditional GCW restrained by means of rigid brackets. In accordance with Fig. 8(b), moreover, maximum reaction forces monitored at the GCW corners are in the order of ≈5 kN for the unit with VE fasteners, compared to ≈38 kN for the module with rigid brackets, with marked benefits for the primary structure. In Figs 8(e)–(f) the distribution of tensile stresses is finally also compared, for the GCW panel with rigid supports or VE fasteners. Worth of interest, in the same figures, is the effect of hard rubber gaskets interposed between the VE compound and the steel bracket (i.e. cross-section of Fig. 3). As far as an infinitely rigid stop is considered only for the deformed VE layer, in fact, an abrupt increase of measured reaction forces at the supports, as well as maximum deflections and stresses in glass can be observed. The same fully rigid stops, in particular, would involve an additional impulse for the sliding GCW unit. As a result, the potential benefits of the same VE fasteners could fully vanish, with a dynamic performance of the GCW panels rather in close correlation with the traditional modules.

5 CONCLUSIONS

In this paper, the feasibility and potential of special mechanical joints interposed at the interface between a given multi-storey primary building and a traditional glazing curtain wall have been investigated via accurate finite-element (FE) numerical models, under various impact scenarios. Careful consideration has been paid for seismic events as well as blast loads, being representative of exceptional and high rise loading conditions for facades and buildings. Based on properly designed fasteners able to introduce additional flexibility and damping capacities in the traditional building, the maximum effects and benefits of such connectors have been emphasized, both in terms of global performance as well as local and component behaviour for the case study object of investigation, giving evidence of the potential of GCWs acting as distributed tuned mass dampers (TMDs) for multi-storey buildings under hazards.

ACKNOWLEDGEMENTS

Part of this research study has been financially supported within the FRA2016 'Inverse' project ('Università degli Studi di Trieste – Finanziamento di Ateneo per progetti di ricerca scientifica – FRA2016'). 'Fondazione CaRiGo' is also acknowledged for providing financial support.

REFERENCES

[1] Larcher, M., Arrigoni, M., Bedon, C., Van Doormaal, A., Haberacker, C., Hüsken, G., Millon, O., Saarenheimo, A., Solomos, G., Thamie, L., Valsamos, G., Williams, A. & Stolz, A., Design of blast-loaded glazing windows and facades: a review of essential requirements towards standardization. *Advances in Civil Engineering*, **2016**, Article ID. 2604232, pp. 14, 2016.
https://doi.org/10.1155/2016/2604232

[2] EN 1998-1-1. *Eurocode 8 – Design of structures for earthquake resistance – Part 1: General rules, seismic actions and rules for buildings*. CEN, Brussels, Belgium, 2004.

[3] FEMA 450. Recommended provisions for seismic regulations for new buildings and other structures, 2003.

[4] ISO 16933:2007/Cor1:2008, Glass in building - Explosion-resistant security glazing - Test and classification for arena air-blast loading, 2008.

[5] GSA-TS01. *Standard test method for glazing and window systems subject to dynamic overpressure loadings*. US General Services Administration, 2003.

[6] Moon, K.S., Vertically distributed multiple tuned mass dampers in tall buildings: performance analysis and preliminary design. *The Structural Design of Tall and Special Buildings*, **19**, pp. 347–366, 2010.
https://doi.org/10.1002/tal.499

[7] Bedon, C. & Amadio, C., Exploratory numerical analysis of two-way straight cable-net facades subjected to air blast loads. *Engineering Structures*, **79**, pp. 276–289, 2014.
https://doi.org/10.1016/j.engstruct.2014.08.023

[8] Amadio, C. & Bedon, C., Viscoelastic spider connectors for the mitigation of cable-supported facades subjected to air blast loading. *Engineering Structures*, **42**, pp. 190–200, 2012.
https://doi.org/10.1016/j.engstruct.2012.04.023

[9] Bedon, C. & Amadio, C., Passive control systems for the blast enhancement of glazing curtain walls under explosive loads. *The Open Civil Engineering Journal*, **10**, pp. 3–26, 2016.

[10] Park, S.W., Analytical modeling of viscoelastic dampers for structural and vibration control. *International Journal of Solids and Structures*, **38**(44–45), pp. 8065–8092, 2000.

[11] Xiaoming, C., Jin, D. & Yungui, L., *Mass proportional damping in nonlinear time-history analysis*. Proceeding of 3rd IC3ME2015, June 27–28, Guangzhou, China, 2015.

[12] Alipour, A. & Zareian, F., *Study Rayleigh damping in structures – Uncertainties and treatments*. Proceedings 14th WCEE, October 12–17, Beijing, China, 2008.

[13] Chopra, A.K., *Dynamics of Structures – Theory and Applications to Earthquake Engineering*, 4th ed., Prentice Hall, Upper Saddle River, New Jersey, 2011.

[14] Systèmes Dassault. ABAQUS v. 6.12, Providence, RI, 2013.

[15] Bedon, C., Amadio, C. & Sinico, A., Numerical and analytical investigation on the dynamic buckling behavior of glass columns under blast. *Engineering Structures*, **79**, pp. 322–340, 2014.
https://doi.org/10.1016/j.engstruct.2014.08.024

HUMAN SAFETY CRITERIA FOR RISK-BASED STRUCTURAL DESIGN

MIROSLAV SYKORA[1], MILAN HOLICKY[1], KAREL JUNG[1] & DIMITRIS DIAMANTIDIS[2]
[1]Czech Technical University in Prague, Klokner Institute, Prague, Czech Republic.
[2]OTH Regensburg, Faculty of Civil Engineering, Regensburg, Germany.

ABSTRACT

Risk and reliability criteria are well established in many industrial sectors such as the offshore, chemical or nuclear industries. Comparative risk thresholds have been specified to allow a responsible organization or regulator to identify activities, which impose an acceptable level of risk concerning the participating individuals, or society as a whole. The scope of this contribution is to present target reliability criteria based on acceptable human safety levels. Application of theoretical principles is illustrated by examples of railway engineering structures. Initially it is shown how civil engineering structures for which human safety criteria play a role are classified according to Eurocodes. Examples include bridges, tunnels or station buildings. The general concepts for risk acceptance are then briefly reviewed, particularly in their relation to the target reliability criteria. The distinction between the two types of criteria is made: group risk and the acceptance criterion based on the Life Quality Index LQI approach introduced by ISO 2394:2015. The differences between the criteria for new and existing structures are discussed. The application is illustrated by an example of a bridge crossing an important railway line. It appears that while benefits and costs of a private stakeholder or public authority are reflected by economic optimisation, the society should define the limits for human safety to achieve uniform risks for various daily-life activities and across different industrial sectors.

Keywords: group risk, human safety, individual risk, Life Quality Index, railway, risk acceptance, structure, target reliability

1 INTRODUCTION

Risk and reliability acceptance criteria are well established in many industrial sectors such as the offshore, chemical or nuclear industries. Comparative risk thresholds have thereby been specified to allow a responsible organization or regulator to identify activities which impose an acceptable level of risk on the participating individuals, or on the society as a whole.

The scope of this contribution is to present target reliability criteria for railway engineering structures based on acceptable human risks. The structures under consideration are initially classified. Civil engineering structures such as bridges, station buildings, tunnels, etc. are taken into account, while electrical and mechanical components such as signalling systems, ventilation systems, electrification systems are not considered herein. The general concepts for risk acceptance are then briefly reviewed, particularly in their relation to the target reliability criteria. The application of theoretical principles is demonstrated by the example of a bridge crossing an important railway line. The difference between new and existing structures is highlighted and recommendations for practical applications are drawn.

2 CLASSIFICATION OF STRUCTURES

Railway civil engineering structures considered herein cover a wide range of constructions and can be divided into:

1. Track carrying structures such as bridges, viaducts, trestles or earth structures.
2. Ancillary structures such as tunnels, stations, towers, platforms or loading docks.

DOI: 10.2495/SAFE-V8-N2-287-298

The aforementioned structures can be classified according to the design working life – for example, 50 years for station buildings and 100 years for bridges and tunnels. They can be also classified according to the possible consequences in case of failure, i.e. minor, medium, large consequences. The second type of classification is for the determination of the target reliability of importance: structures with large potential consequences are associated with higher target reliability.

A classification of civil engineering structures according to various normative documents was provided by Sykora *et al.* [1]. As an example, UIC 7772 [2] makes distinction between:

1. Superstructures supporting elevated structures that are permanently occupied (offices, lodgings, business premises), structures for short-term gathering of people (theatres, cinemas) and multi-storey structures subject to short-term occupancy (multi-storey car parks, warehouses)
2. Superstructures not supporting elevated structures (roadways, road bridges, railway bridges, footbridges and similar structures) and single-storey structures not providing long-term occupancy (parking areas, warehouses).

In this contribution, the failure consequence classes according to EN 1990 [3] for basis of structural design are adopted:

- Low consequence for loss of human life, *and* economic, social or environmental consequences small or negligible; examples based on authors' experience: low rise buildings where only few people are present or small supply structures.
- Medium consequence for loss of human life, economic, social or environmental consequences considerable; examples: minor bridges.
- High consequence for loss of human life, *or* economic, social or environmental consequences very great; examples: normal and major bridges, and tunnels, major railway stations.

3 RISK ACCEPTANCE CRITERIA

Explicit risk acceptance criteria are commonly applied in many industrial sectors in order to provide either a quantitative decision tool for the regulator, or a comparable requirement for the industry when dealing with the certification/approval of a particular structure or system. The following criteria are commonly applied and analytically described [4, 5]:

1. *Individual risk criteria*: no individual (or group of individuals) involved in a particular activity can be exposed to an 'unacceptable' risk; a typical value is 10^{-6} per year [6]. If an individual worker or a member of the public is found to be exposed to excessive risk, safety measures are adopted regardless of the cost-benefit effectiveness.
2. *Group risk criteria:* a certain activity must not produce high frequency occurrences of large-scale accidents (i.e. with particularly severe consequences). This means that the 'unacceptable" level of risk varies for different accident magnitudes. This principle attempts to capture a supposed socio-political aversion to large accidents and provides a regulatory basis (i.e. enforced investments in safety) in situations where the other criteria do not call for intervention.

The requirements based on criterion (1) can be significantly affected by the relative time fraction for which a person occupies or uses a structure. For railway civil engineering structures, it is assumed that this fraction is commonly low (exceptions may include most exposed workers), individual risk criteria become less important and group risk criteria dominate the derivation of target reliability values [7–9]. These criteria include human, economic and environmental criteria and are briefly reviewed in the following sections.

4 GROUP RISK

The group risk is often represented in the form of a numerical *F-N*-curve where *N* represents the number of fatalities and *F* is the frequency of accidents with more than *N* fatalities [4, 5]. This curve shows the probability of exceedance as a function of the number of fatalities *N*, commonly using a double logarithmic scale [10]:

$$1 - F_N\left(x\right) = \mathrm{P}\left(N > x\right) = \int_x^{\infty} \mathrm{f}_N\left(\xi\right)\mathrm{d}\xi \tag{1}$$

where $\mathrm{f}_N(x)$ = probability density function of number of fatalities per year; and $\mathrm{F}_N(x)$ = cumulative density function, the value of which gives a probability of $\leq x$ fatalities per year. A simple measure for group risk is the annual expected value of the number of fatalities, which is frequently used to compare alternative projects in terms of their inherent risk with respect to human safety.

Typical *F–N* curves reported in the literature show different patterns for the same industrial activity in various countries or for different industrial activities in the same country. The following general formula has been proposed to represent the group risk acceptance criterion:

$$F \leq a\, N^{-k} \tag{2}$$

where a and k = predefined constants that can be related to statistical observations from natural and man-made hazards [11]. Some natural hazards show relationships with k slightly smaller than unity, while most manmade hazards are described by a relationship with $k > 1$. From statistical observations the constants a and k vary widely depending on the type of hazard and the type of technical activity. It was proposed to set the constants in such a way that the curve envelops the curves for most natural hazards and some of the more common manmade hazards [7]. For acceptable risks related to structural failures, the constant would be around $a = 10^{-6}$ and for marginally acceptable risks $a = 10^{-4}$; $k = 1$ represents risk-neutral curves, $k > 1$ describes curves with risk aversion and $k < 1$ curves with risk proneness. The case of $k < 1$ leads to infinitely large expected losses (in terms of lives or cost) and, therefore, is not acceptable.

The constant a represents the frequency of occurrence of events with one and more fatalities; commonly, annual values are considered. Its value should be consistent with the reference system to which eqn (2) is applied. The reference system can range from a group of structures to an individual structural member and can include other structure-specific and industry-specific parameters [12]. For railway infrastructures, a national scale is assumed [5, 12, 13].

Based on the *F-N* curves, the so-called ALARP – as low as reasonably possible – region can be defined by two limits [5]. The area above upper limit represents the risk that is not tolerated in any circumstances, while the risk below the lower limit is of no practical interest. Such acceptability curves have been developed in various industrial fields, including the chemical and the transportation industries.

In the ALARP principle the 'width' between the upper and lower bound curves is of importance. This width is often two orders of magnitudes allowing for excessive flexibility in practical cases. Examples of fatality criteria (*F–N* curves) given in Fig. 1 are based on the recommendations provided in:

- ISO 2394:1998 for general principles of structural reliability (superseded in 2015): $a = 10^{-2}$ per annum and $k = 2$ – examples provided as a risk acceptance criterion in structural design to avoid accidents where large numbers of people may be killed, deemed to be associated with a collapse of the whole structure
- Trbojevic [14]: $a = 2 \times 10^{-5}$ per annum and $k = 1$ – criterion for a structure with 100 persons at risk
- *fib* bulletin 80 [9]: $a = 5 \times 10^{-4}$ per annum and $k = 2$ – criterion derived for a main load-bearing member of a road bridge in order to maintain safety levels associated with current best practice
- ISO 2394:2015 [15], the LQI approach – see Section 5.

Note that the ISO and *fib* COM3 recommendations are based on the notion that *N* represents the expected number of fatalities in an event and thus a limiting probability does not have the cumulative character of eqn (2). However, the numerical difference between these concepts is commonly negligible due to small probabilities under consideration. This is why the ISO and *fib* COM3 curves are hereafter considered as directly comparable with *F–N* curves. The ISO recommendation is applied in some countries; for instance Belgium and the Netherlands consider the acceptability criterion for the events with $N \geq 10$ fatalities [6].

The criteria in Fig. (1a) provide values ranging across more than two orders of magnitude. This is attributed to different reference systems for which these criteria are applicable. Whilst the example in ISO 2394:1998 is considered to be used for large groups of structures, the other two recommendations are focused on individual buildings [14], or on a key structural member of a bridge [9]. Hence, the reference system is of key importance when establishing the criteria for human safety.

Figure (1b) displays annual reliability indices β derived from probability $F(N)$ given in Fig. (1a); see EN 1990 [3]. It appears that the criteria under consideration lead to a wide range of β-values particularly for events with $N < 10$.

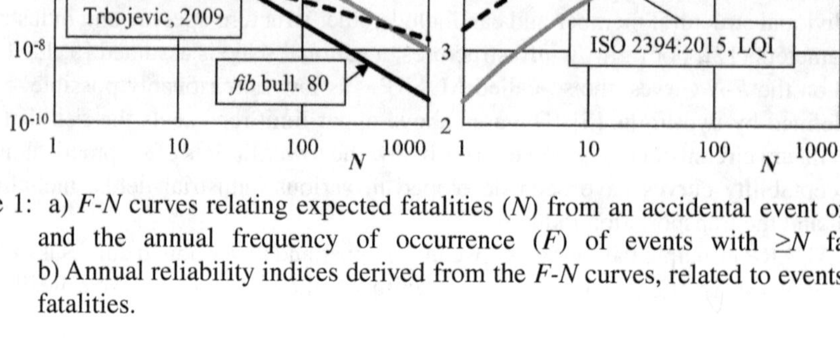

Figure 1: a) *F-N* curves relating expected fatalities (*N*) from an accidental event or failure and the annual frequency of occurrence (*F*) of events with ≥*N* fatalities; b) Annual reliability indices derived from the *F-N* curves, related to events with *N* fatalities.

It is noted that human safety does not only involve fatalities but also injuries. In many studies injuries are related to fatalities by using a multiplicative factor, for example 0.1 for moderate injury and 0.5 for major injury. Based on this simple procedure weighted fatalities can be obtained. For detailed discussion see Ref. [16].

Diamantidis *et al.* [17] derived human risk acceptance criteria for the railway industry by assuming that the safety inherent in the traditional railways in the last two or three decades is acceptable. The safety target was therefore derived by analysing recent risk history of the railway system in terms of frequency of occurrence of accidents and the extent of their consequences. The procedure generally used to estimate the risk associated with railway transport is based on the analyses of the frequency of occurrence of given consequences for a given accident. The following classification of consequences was proposed:

- Medium consequences: 1 to 10 fatalities with an average value of 3 fatalities;
- Severe consequences: 11 to 100 fatalities with an average value of 30 fatalities;
- Catastrophic consequences: 101 to 1000 fatalities with an average value of 300 fatalities.

The evaluation of the probability can be performed by assuming that accidents occur according to a Poisson process [18]. In general a Bayesian approach by which expert judgements can be combined with statistical data is recommended, due to the limited experience with accidents. This approach was applied to data of recorded accidents of the Italian, Austrian and German railways [17]. The obtained results are considered valid for a first definition of an acceptable safety level for the Western European railway systems, and are comparable to computed values for various tunnel projects. The results and the recommended acceptance criteria can be summarized as follows:

- Events of medium consequences are associated an annual probability of 10^{-9} per train-kilometre;
- Events of severe consequences are associated an annual probability of 10^{-10} per train-kilometre;
- Events of catastrophic consequences are associated a probability of 10^{-11} per train-kilometre.

Example: Consider a railway tunnel with a length of 50 kilometres and a daily traffic flow of 200 trains in both directions. The acceptable return periods associated with accidental events are derived from the aforementioned results as follows:

- Number of train-kilometres per year: $50 \times 200 \times 365 = 3.65 \times 10^6$;
- Accidents associated with medium consequences: acceptable annual probability of an accident $p \leq 3.65 \times 10^6 \times 10^{-9} = 0.00365$; acceptable return period $T \geq 1 / p \approx 300$ years;
- Severe consequences: 3000 years;
- Catastrophic consequences: 30000 years.

The derived acceptable return period for medium consequences is approximately within the range of the working life of important infrastructures, such as a long tunnel. For catastrophic consequences, the return period is of the same order of magnitude as is considered when designing industrial plants.

It is noted that, along with human risks, economic and environmental risks play an important role in decision making. Economic losses are direct consequences related, for example, to the repair of initial damage, replacement of structure and equipment, and indirect consequences such as loss of production, temporary relocation, rescue costs or loss of reputation. Environmental consequences can be presented in terms of permanent or long-term damage to terrestrial, freshwater, marine habitats and groundwater reservoirs. For more details, see Refs. [1, 19].

5 LIFE QUALITY INDEX

5.1 ISO 2394 recommendations

The Life Quality Index LQI, as for example discussed with respect to its implementation in technical standards in [20], was developed to support decisions related to allocations of available public resources between and within various societal sectors and industries. According to ISO 2394:2015 [15], the LQI is an indicator of the societal preference and capacity for investments into life safety expressed as a function of GDP, life expectancy at birth and ratio between leisure to working time.

The ISO standard [15] provides the detailed guidance on how preferences of the society regarding investments into health and life safety improvements can be described by the LQI concept. The target level is derived by considering the costs of safety measures, the monetary equivalent of societal willingness to save one life, and the expected number of fatalities in the event of structural failure. Essentially, this approach combines economic and human safety aspects. Compared with economic optimization [8, 21–23], it should lead to lower target reliability indices, as only the human consequences of structural failure are taken into account, while other losses such as economic and environmental costs are not taken into account. In the LQI approach, the danger to which the people are subjected might vary on an individual basis within the group of people affected, which may be deemed unethical [8].

The tentative minimum annual target reliabilities provided in ISO 2394 [15] are based on the following acceptance criterion:

$$-\frac{\mathrm{d}\,p_f(d)}{\mathrm{d}\,d} \le K_l = \frac{C_1(d) \times (\gamma_s + \omega)}{SWTP \times N} \tag{3}$$

where p_f = annual failure probability; d = decision parameter (see the text below); the parameters indicated in Annex G of the ISO standard are adopted in the following analysis:

- $K_1 = 10^{-4.5}$ for small relative life-saving cost, deemed relevant for most design situations where higher reliability can be readily achieved;
- $\gamma_s = 0.03$ as an interest rate of relevance for decision-making on behalf of society, moderately selected rate of economic growth;
- $\omega = 0.02$ as an annual rate of obsolescence that approximately corresponds here to the reciprocal value of a lifetime;
- Societal willingness to pay $SWTP$ = 5 100 000 CHF;
- $N = 10$ is the expected number of fatalities that could be associated with Consequence Class CC3 according to ISO 2394 [15] where CC3 is associated with $N < 50$ and CC2 with $N < 5$; five CCs are distinguished in the ISO standard. Medium failure consequences may be related to ten fatalities in [14]. Note that the empirical data in [9, 12] provide indications on the relationship between N and collapsed floor area (buildings) or span length (bridges).

For these notional values, marginal costs associated with a considered safety measure obtained from eqn (3) become $C_1(d)$ = 32 000 CHF and ISO 2394 [15] indicates annual reliability index of 4.2, which corresponds to the maximum acceptable failure rate $F(N = 10) = 1.3 \times 10^{-5}$ per year (Fig. 1).

5.2 Application to a railway bridge

To provide a better insight into the performance of criterion (3), LQI reliability levels are derived for a key load-bearing member of a railway bridge following the simple example provided in Annex G of ISO 2394 [15]. A fundamental limit state function is considered:

$$Z(d) = R(d) - E \tag{4}$$

where R = resistance; E = annual maxima of a load effect; and d = decision parameter defined here as the ratio between the mean values of R and E, $d = \mu_R / \mu_E$. The annual failure probability is obtained as $p_f(d) = P[Z(d) < 0]$.

The following probabilistic models of basic variables are taken into account:

- Lognormal distribution with coefficient of variation of 0.2 for resistance covering uncertainties in material properties, geometry and resistance model uncertainty related to flexural or shear resistance, axial compression, buckling etc.
- Gumbel distribution with coefficient of variation of 0.4 for annual maxima of a combined load effect covering:
  - permanent actions
  - time-invariant components of variable actions such as shape, exposure and other factors or load model simplifications
  - annual maxima of time-variant components of variable actions, for instance snow load on the ground, basic wind velocity pressure, imposed or traffic loads.

The aforementioned notional values of the input parameters of eqn (3), deemed to describe the alternative of “small relative life-saving costs” according to the ISO standard, are adapted to the railway bridge example as follows:

- ω is decreased to 0.01, corresponding to a standard lifetime of bridges of 100 years;
- γ_s is reduced to 0.0285 following the guidance of the Green Book of the UK Treasury for public sector projects.

while *SWTP* and $C_1(d)$ are considered by the same values and N is a study parameter.

Figure 2 shows annual reliability indices related to events with N fatalities based on the LQI approach in ISO 2394 [15] and, for the sake of comparison, the criterion for human safety for bridges in *fib* bulletin 80 [9]. It is observed that for larger N, the *fib* bulletin 80 for existing structures requires higher reliability levels than the ISO standard for new structures (‘small relative life-saving costs’). This is in contradiction with numerous previous studies advocating lower target levels for existing structures [8, 19, 23–25] and with the theoretically justified approach to reduce target reliability levels for increasing costs of safety measures. The discrepancy is primarily attributable to the risk-averse value $k = 2$ adopted in the *fib* bulletin. When a risk-neutral unity is taken into account, the *fib* curve in Fig. 2 becomes parallel and close to the ISO level for medium relative life-saving costs.

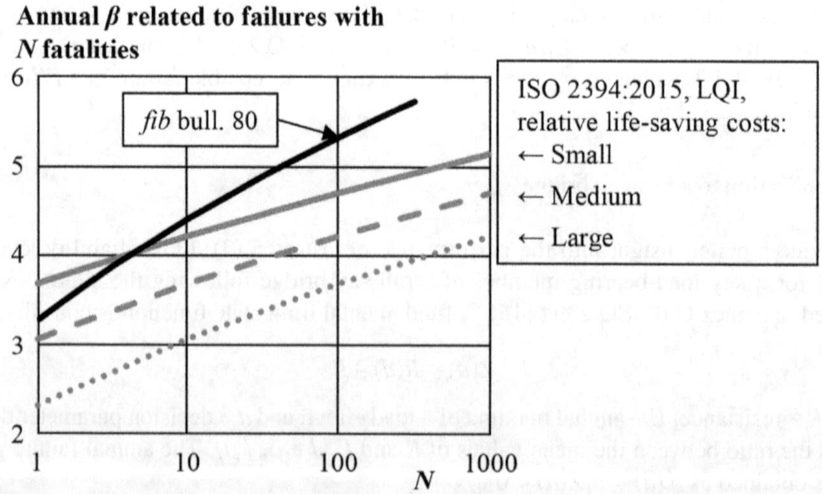

Figure 2: Annual reliability indices β related to events with $\geq N$ fatalities, based on the LQI approach in ISO 2394 [15] and the criterion for human safety for existing bridges in *fib* bulletin 80 [9].

To indicate appropriate target reliability for existing structures, LQI β-values are further derived for 'medium' and 'large' relative life-saving costs. Following the example in Annex G of ISO 2394 [15], this can be achieved by increasing $C_1(d)$ by one or two orders of magnitude for medium and large relative life-saving costs, respectively [16]. This results in decreasing β-levels (Fig. 2):

- By about 0.5 for medium relative life-saving costs
- By more than 1.0 for large relative life-saving costs.

More information on the LQI approach including application in the bridge industry is provided in [26]. The differences in target reliability can be readily reflected by updating partial factors in practical verifications [27, 28].

6 CASE STUDY

Practical application of the human safety criteria is demonstrated by the example of a bridge pier located in the vicinity of a railway line (Fig. 3) and the related hazard of accidental impact of a train on the pier. A probabilistic risk analysis was employed to optimise the bridge design from the perspective of an owner of the bridge. The analysis described in Ref. [1] was based on an event tree approach, considering multiple hazards scenarios related to derailment of a train, impact to the bridge, collapse of the bridge due to impact and secondary collision with a train in the opposite direction. The UIC recommendations related to probabilities of unfavourable events [2] along with real traffic were taken into account. Figure 4 displays the computed F–N curves for different train speeds. As an example for a train speed of 120 km/h, the curve was obtained as follows:

- With an annual probability of 1.3×10^{-7} (equal to frequency due to its very low value), the most adverse scenario – derailment of the train and impact to the bridge, collapse of the

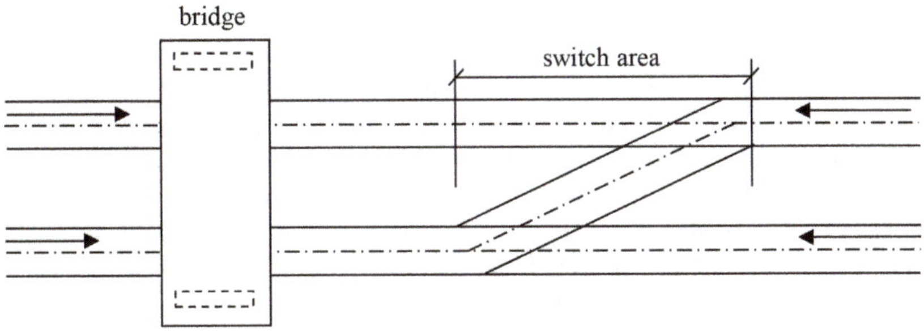

Figure 3: Plan view of the bridge.

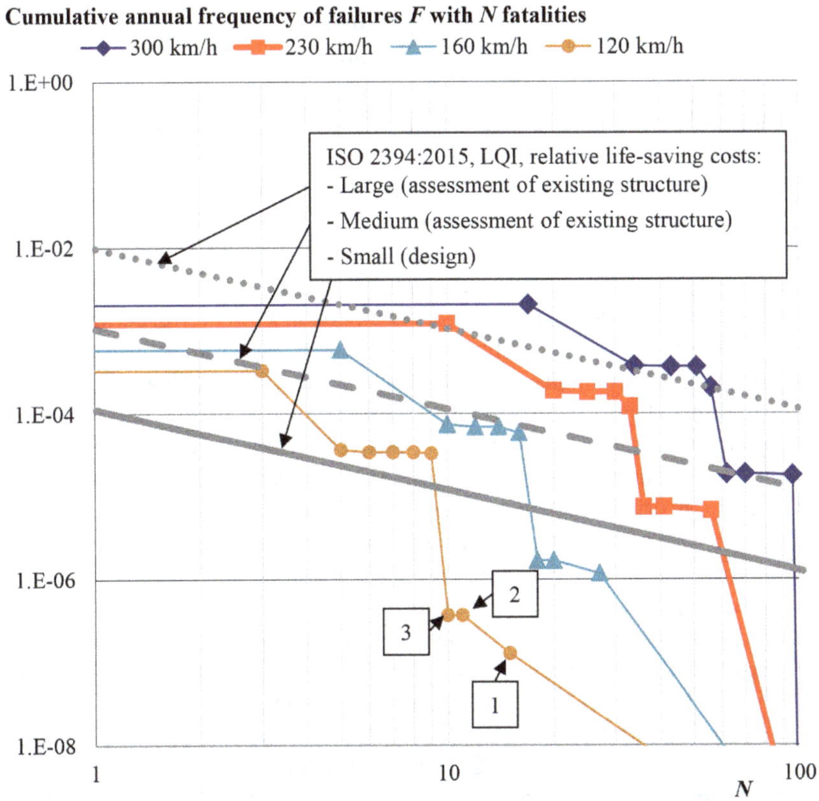

Figure 4: *F–N* curves for different train speeds and the LQI acceptance criteria according to ISO 2394 [15].

bridge and secondary collision with a passenger train in the opposite direction – leads to 15 fatalities (point 1 in Fig. 4).

- With an annual probability of 2.4×10^{-7}, the second most adverse scenario – derailment, impact, secondary collision with a passenger train, but no bridge collapse – results in 11 fatalities. At least 11 fatalities are then associated with a cumulative frequency F of $1.3 \times 10^{-7} + 2.4 \times 10^{-7} = 3.7 \times 10^{-7}$ (point 2).
- With negligible probability, the third most adverse scenario yields 10 fatalities and in Fig. 4 is thus associated with a cumulative frequency F of 3.7×10^{-7} (point 3).

It remains to be verified whether the optimum design strategy complies with the criteria on human safety that may be postulated by public authorities. As an example the LQI acceptance criteria according to ISO 2394 [15] are plotted in Fig. 4. It is shown that for any train speed under consideration, the human safety criteria suggest that human risks exceed the commonly acceptable level for design situation and further design measures to mitigate human safety risks are required. Options include increasing the lateral distance of the bridge piers beyond the optimum value, or increasing the distance between the bridge and switch area (Fig. 3).

When an existing bridge is considered and a possibility of the change of the switch area is disregarded, increasing the lateral distance of the bridge piers from the railway tracks is an excessively costly measure. The acceptance criterion for large relative life-saving costs may then be applied and a train speed of 230 km/h can be approved (Fig. 4).

7 CONCLUSIONS

The following conclusions can be drawn from the presented study:

- The human safety criteria should always be applied to check whether or not the design strategy based on economic optimisation including compensation costs related to fatalities, injuries and environmental impacts leads to acceptable risks to persons endangered by structural failure.
- While benefits and costs of a private stakeholder or public authority are reflected by economic optimisation, the society defines the limits for human safety through: (1) The individual risk criterion that provides comparison with risks related to other activities that are acceptable for the society. This criterion is particularly useful when risks vary across the range of persons potentially dependent on structural failure. However, the criterion is significantly affected by the relative time fraction for which a person occupies or uses a structure. As this fraction is very low for transportation structures, the criterion is commonly not decisive for bridges or tunnels. It can be used to assess risks of most exposed workers.
  (2) The group risk criteria, often expressed by *F–N* curves, lack a sound theoretical basis and do not consider costs of safety measures; they are however applied in many industrial sectors in which major accidents may occur. The major difficulty is to define a reference system for their application.
- The LQI criterion takes into the cost efficiency of changing design parameters and of the safety measures. However, this criterion is dependent on numerous uncertain input parameters and it may be difficult to provide general practical recommendations.

ACKNOWLEDGEMENTS

This work was supported by the Czech Science Foundation under Grants P105/12/G059 and 16-11378S, by the Regensburg Center of Energy and Resources (RCER) of OTH Regensburg and by the Technology- and Science Network Oberpfalz (TWO). Outcomes of COST Action TU1402 have been utilized.

NOTE

1 *ORCID: http://orcid.org/0000-0001-9346-3204*

REFERENCES

[1] Sykora, M., Holicky, M., Jung, K. & Diamantidis, D., *Railway RAM and safety handbook (Chapter 45: Target reliability for new and existing railway civil engineering structures, accepted for publication)*, CRC Press/ Balkema: Leiden, 2017.

[2] UIC 777-2, *UIC Code 777-2, Structures built over railway lines, Construction requirements in the track zone,* International Union of Railways: 2002.

[3] EN 1990, *Eurocode - Basis of structural design,* CEN: Brussels, pp. 87, 2002.

[4] HSE, *Reducing risks, protecting people,* Health & Safety Executive: Norwich (UK), pp. 88, 2001.

[5] CIB TG 32, *Risk assessment and risk communication in civil engineering (Report 259),* CIB: Rotterdam, 2001.

[6] DNV GL, *Harmonised risk acceptance criteria for transport of dangerous goods,* European Commission DG-MOVE: pp. 183, 2014.

[7] JCSS, *Probabilistic assessment of existing structures.* Joint Committee on Structural Safety, edited by D. Diamantidis, RILEM Publications S.A.R.L., 2001.

[8] Steenbergen, R.D.J.M., Sykora, M., Diamantidis, D., Holicky, M. & Vrouwenvelder, A.C.W.M., Economic and human safety reliability levels for existing structures. *Structural Concrete,* **16**, pp. 323–332, 2015.
https://doi.org/10.1002/suco.201500022

[9] *fib* COM3 TG3.1, *Partial factor methods for existing structures (fib bulletin 80, recommendation), fib*: pp. 129, 2016.

[10] Jonkman, S.N., van Gelder, P.H.A.J.M. & Vrijling, J.K., An overview of quantitative risk measures for loss of life and economic damage. *Journal of Hazardous Materials,* **99**(1), pp. 1–30, 2003.
https://doi.org/10.1016/s0304-3894(02)00283-2

[11] Vrouwenvelder, T., Leira, B.J. & Sykora, M., Modelling of hazards. *Structural Engineering International,* **22**(1), pp. 73–78, 2012.
https://doi.org/10.2749/101686612x13216060213356

[12] Tanner, P. & Hingorani, R., Acceptable risks to persons associated with building structures. *Structural Concrete,* **16**(3), pp. 314–322, 2015.
https://doi.org/10.1002/suco.201500012

[13] Vrijling, J., van Gelder, P. & Ouwerkerk, S., Criteria for acceptable risk in the Netherlands. *Infrastructure Risk Management Processes, American Society of Civil Engineers,* pp. 143–157, 2005.
https://doi.org/10.1061/9780784408155.ch05

[14] Trbojevic, V.M., Another look at risk and structural reliability criteria. *Structural Safety,* **31**(3), pp. 245–250, 2009.
https://doi.org/10.1016/j.strusafe.2008.06.019

[15] ISO 2394, *General principles on reliability for structures.* ISO: Geneve, Switzerland, pp. 111, 2015.

[16] Diamantidis, D., Holicky, M. & Sykora, M., Target reliability levels based on societal, economic and environmental consequences of structural failure (accepted for publication). *Proc. ICOSSAR 2017,* CRC Press/Balkema: Leiden (The Netherlands), pp. 10, 2017.

[17] Diamantidis, D., Zuccarelli, F. & Westhäuser, A., Safety of long railway tunnels. *Reliability Engineering & System Safety,* **67**(2), pp. 135–145, 2000.
https://doi.org/10.1016/s0951-8320(99)00059-9

[18] Melchers, R.E., *Structural reliability analysis and prediction,* John Wiley & Sons Ltd.: Chichester, England, pp. 437, 2001.

[19] Sykora, M., Holicky, M., Jung, K. & Diamantidis, D., Target reliability for railway civil engineering structures (accepted for publication). *Proc. ESREL 2017,* Taylor and Francis/ Balkema: Leiden, 2017.

[20] Rackwitz, R., Optimization - the basis of code-making and reliability verification. *Structural Safety,* **22**(1), pp. 27–60, 2000.
https://doi.org/10.1016/s0167-4730(99)00037-5
[21] Holicky, M., Optimisation of the target reliability for temporary structures. *Civil Engineering and Environmental Systems,* **30**(2), pp. 87–96, 2013.
https://doi.org/10.1080/10286608.2012.733373
[22] Holicky, M., Diamantidis, D. & Sykora, M., Determination of target safety for structures. *Proceeding of ICASP12,* eds. T. Haukaas, pp. 8, 2015.
[23] Sykora, M., Holicky, M., Jung, K. & Diamantidis, D., Target reliability for existing structures considering economic and societal aspects. *Structure and Infrastructure Engineering,* **13**(1), pp. 181–194, 2016.
https://doi.org/10.1080/15732479.2016.1198394
[24] Steenbergen, R.D.J.M. & Vrouwenvelder, A.C.W.M., Safety philosophy for existing structures and partial factors for traffic loads on bridges. *Heron,* **55**(2), pp. 123–139, 2010.
[25] Holicky, M., Markova, J. & Sykora, M., Target reliability levels in present standards. *Transactions of the VSB - Technical University of Ostrava, Civil Engineering Series,* **14**(2), pp. 46–53, 2014.
[26] Fischer, K., Virguez, E., Sánchez-Silva, M. & Faber, M.H., On the assessment of marginal life saving costs for risk acceptance criteria. *Structural Safety,* **44**, pp. 37–46, 2013.
https://doi.org/10.1016/j.strusafe.2013.05.001
[27] Sykora, M., Holicky, M. & Markova, J., Verification of existing reinforced concrete bridges using the semi-probabilistic approach. *Engineering Structures,* **56**, pp. 1419–1426, 2013.
https://doi.org/10.1016/j.engstruct.2013.07.015
[28] Caspeele, R., Sykora, M., Allaix, D.L. & Steenbergen, R., The design value method and adjusted partial factor approach for existing structures. *Structural Engineering International,* **23**(4), pp. 386–393, 2013.
https://doi.org/10.2749/101686613x13627347100194

OPTIMAL TRADEOFFS BETWEEN THE SECURITY AND COST OF CRITICAL BUILDINGS AND INFRASTRUCTURE SYSTEMS

STEVEN SCHULDT & KHALED EL-RAYES
Department of Civil and Environmental Engineering, University of Illinois at Urbana-Champaign, Urbana, IL, USA.

ABSTRACT

Explosive terrorist attacks targeting critical buildings and infrastructure systems pose a formidable threat worldwide, having caused 12,425 casualties and $20 billion in direct economic losses in 2015 alone. Designers of these critical buildings attempt to minimize the security risks to site personnel and buildings by analyzing and selecting the most effective combination of: (1) increasing the standoff distance between site assets and potential locations of explosive attacks; (2) constructing blast-mitigating perimeter walls; and (3) hardening site facilities. To support designers in this critical and challenging task, this paper presents the development of a multi-objective optimization model capable of generating optimal tradeoffs between minimizing total site destruction levels and minimizing site construction costs. The model computations are performed utilizing the nondominated sorting genetic algorithm II (NSGA-II) because of its proven capability in modeling non-linear objective functions and constraints, and its successful modeling of previous facility layout problems. The model performance was evaluated using a case study of a hypothetical military forward operating base, and the results illustrated the novel capabilities of the developed model in identifying design configurations that generate optimal tradeoffs between the aforementioned optimization objectives. These capabilities are expected to support designers in their ongoing efforts to construct cost-effective sites that minimize the security risks to personnel and buildings from the threat of explosive terrorist attacks.

Keywords: blast effects, blast wall, critical infrastructure, facility layout, genetic algorithms, optimization, security

1 INTRODUCTION

Explosive terrorist attacks targeting critical buildings and infrastructure systems pose a formidable threat worldwide, having caused 12,425 casualties and $20 billion in direct economic losses in 2015 alone [1,2]. Accordingly, developing optimal site layout and security plans for critical buildings and infrastructure projects is essential to minimize the destruction caused by an explosive terrorist attack. The most cost-effective design solution to minimize building destruction levels is to increase the standoff distance between site assets and the closest possible explosive location [3]. When the available standoff distance does not provide the desired level of protection for the anticipated explosive threat, designers must utilize a combination of constructing blast-mitigating perimeter walls and hardening site buildings [4]. Designers of critical building and infrastructure projects face the challenging task of selecting the optimal design configuration that provides the desired level of protection while adhering to a limited construction budget in a confined site space. Therefore, designers must carefully analyze all feasible design alternatives and select the optimal combination of perimeter blast walls, building locations, and building materials that minimizes site destruction levels from an explosive attack while minimizing site construction costs.

Previous models and research studies evaluated the blast-mitigating capabilities of various blast wall materials [5–10] and attempted to optimize the layout of site facilities to improve site security [11–13]. Despite their contributions, these studies do not provide the capability to optimize the layout of site buildings and the selection of perimeter blast walls and building

DOI: 10.2495/SAFE-V8-N2-299-306

materials in order to minimize total site destruction levels from an explosive attack. Accordingly, the main objective of this study is to present the development of a new multi-objective optimization model capable of generating site layout and security planning design solutions that represent optimal tradeoffs between minimizing total site destruction levels from an explosive attack and minimizing site construction costs.

The following sections of the paper describe: (i) formulating the multi-objective optimization model; (ii) identifying the required model input data; and (iii) evaluating model performance on an application example.

2 MULTI-OBJECTIVE OPTIMIZATION MODEL FORMULATION

The present model utilizes the nondominated sorting genetic algorithm II (NSGA-II) [14] to optimize the site layout and security decisions for critical building and infrastructure systems to minimize total site destruction levels while minimizing site construction costs. The development of this model includes identifying model decision variables, formulating the destruction and cost objective functions, and integrating facility layout constraints.

2.1 Model decision variables

The decision variables utilized in the present model are designed to represent the site layout and security decisions that have an impact on site destruction levels suffered from an explosive attack and site construction costs. Accordingly, the model considers two main types of decision variables: (1) perimeter blast wall decision variables; and (2) building decision variables. First, perimeter decision variables consist of the perimeter blast wall location (PL), blast wall type (T) and blast wall height (H). Second, building decision variables include the selected building material (M_i), building location (Loc_i), and building orientation (θ_i), as shown in Fig. 1.

The model employs a user-specified grid system that allows planners to set their preferred grid spacing. Building locations are then defined by the placement of their centroids on the grid system and the building orientation is the degree that the building is rotated about its centroid. Building location 1 (grid coordinate 44) in Fig. 1 shows two possible building rotation angles for a single building, where the blue building represents a 0° orientation and the orange building represents a 90° orientation.

2.2 Model objective functions

The first objective function in the present model is designed to minimize the destruction of site buildings caused by an explosive attack. This objective function is quantified as the site destruction level (DL) index that ranges from 0%, which represents no building destruction, to 100%, which represents the complete destruction of all buildings on site, as shown in eqn (1). DL is represented as a weighted average of the destruction level suffered by all buildings on site and is calculated by summing up all the products of the damage level of each building and its importance weight (w_i), which is assigned by designers to designate buildings that are most critical to the site mission. The damage level of each building is calculated using a recently developed model for quantifying blast effects on facilities behind blast walls [15]. Beginning at the blast center, this developed methodology computes the distances from the blast location at which five decreasing levels of building destruction occur (severe, heavy, moderate, minor, and minimal) for each unique design combination of blast charge weight,

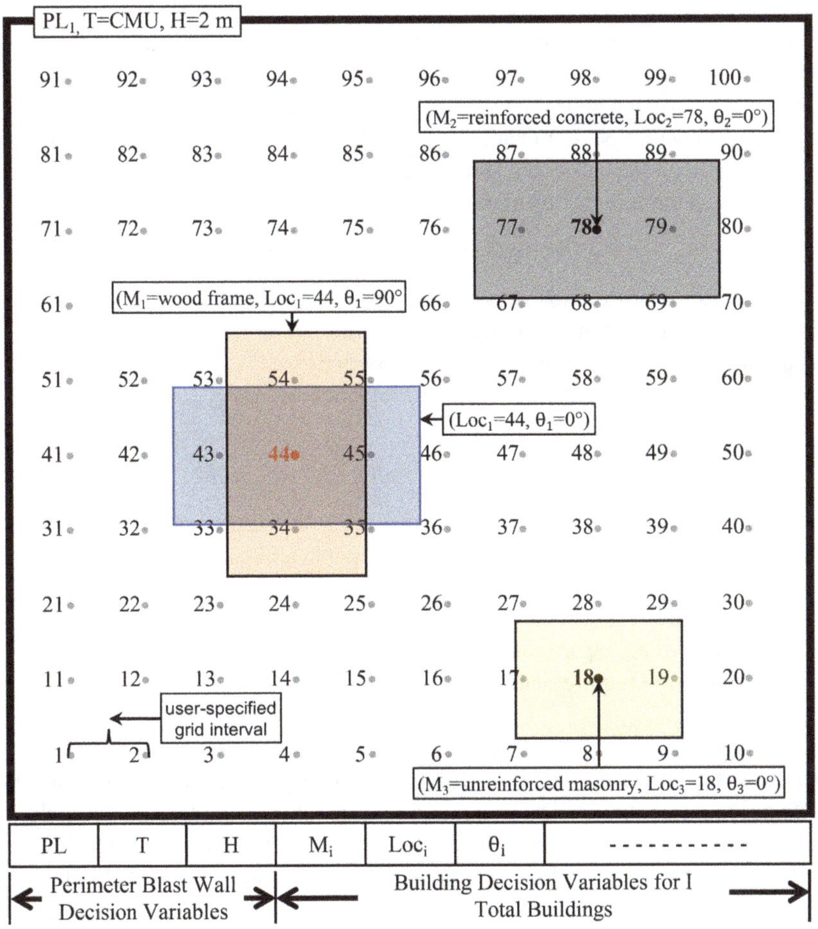

Figure 1: Model decision variables.

perimeter blast wall location, type, and height, and building material. Next, the model automatically computes the percent area of each building within each of these five damage levels ($PBA_{i,j}$). The building damage level is then computed by taking the sum of the $PBA_{i,j}$ multiplied by the level of building destruction associated with each damage level (PD_j), as defined by existing blast design manuals [4,16].

$$minimize\, DL = \sum_{i=1}^{I} \sum_{j=1}^{5} PBA_{i,j} \times PD_j \times w_i \tag{1}$$

where DL = total site destruction level (0–100%);
i = building number;
I = total number of buildings on site;
j = damage level;
$PBA_{i,j}$ = percent area of building i within damage level j (%);
PD_j = percent building destruction caused by damage level j (%); and
w_i = importance weight for each building i, where $w_i > 0$ and $\sum_{i=1}^{I} w_i = 1$.

The second optimization objective is minimizing site construction costs (SC). SC is calculated as the sum of building construction costs (BC), perimeter blast wall construction costs (WC) and future wall replacement costs (WR), as shown in eqn (2). It is necessary to include wall replacement costs when the site operational lifespan is longer than the expected design life of the selected wall type. For example, sand-filled containers typically have a design life of 3–5 years. Furthermore, a discount factor is applied to the wall replacement costs to compute the present value of the future construction costs.

$$minimize SC = BC + WC + WR \tag{2}$$

where SC = total site construction cost (present year $);
BC = construction cost of all buildings on site ($);
WC = perimeter blast wall construction cost ($); and
WR = present value of future year wall replacement cost ($).

2.3 Model constraints

The model is designed to comply with all practical geometric constraints to guarantee the feasibility of generated layout solutions by ensuring: (1) all buildings are constructed within the site perimeter; (2) buildings do no overlap one another; (3) minimum separation distances between buildings are established to comply with applicable security and safety requirements; and (4) maximum separation distances between buildings are established to comply with applicable mission requirements.

3 MODEL INPUT DATA

Designers of critical buildings and infrastructure systems projects are required to collect and provide all building, perimeter blast wall, site, and explosive threat data. Additionally, designers must specify all search parameters for the multi-objective genetic algorithm. Building data defines: (1) dimensions of building(s) footprint; (2) designer-specified building importance weights; (3) available building materials; and (4) unit construction cost for each building material ($/m^2). Perimeter blast wall data includes: (1) feasible location(s) for the perimeter blast wall; (2) available blast wall types; (3) optional heights for each blast wall type; (5) design life for each blast wall type; (6) unit construction cost of each wall type ($/m); and (7) unit cost to remove and replace each blast wall type ($/m). Site data consist of the: (1) dimensions of the available site area; (2) anticipated site operational lifespan; (3) construction area cost factor based on the site location; and (4) discount factor to present all costs in present-year dollars. Explosive threat data define the explosive charge weight(s) to be evaluated and should be selected based upon local intelligence and any historical attack data for the project area. The genetic algorithm search parameters define the population size, number of generations, mutation rate, crossover type and crossover rate. For site layout optimization models using NSGA-II, the following parameters are recommended to provide near-optimal results with practical computational times: population size ranging from 5 to 10 times the total number of decision variables; number of generations approximately 10 times the number of decision variables; mutation rate between 0.001 and 0.1, multi-point crossover, and a crossover rate between 0.5 and 0.9 [17].

4 APPLICATION EXAMPLE

The performance of the developed site layout and security planning optimization model is evaluated using an application example of the construction of a 100-person military forward operating base. The input data required for this example includes: (1) building data; (2) perimeter blast wall data; (3) site characteristics; (4) threat data; and (5) genetic algorithm search parameters. First, building data includes the dimensions and importance weights for the 11 buildings to be constructed on site, as shown in Table 1, as well as the construction cost per square meter for each of the six available building materials. Second, perimeter blast wall data consists of the height, design life, and the initial construction and replacement costs per linear meter for 15 feasible blast wall types. Third, site characteristics include an available site area of 90 meters x 60 meters, an anticipated site operational lifespan of 30 years, and a construction area cost factor of 1.55 based on the site location [18]. Fourth, the application example analyzed a 250 kg explosive threat, which represents the design blast charge weight carried in a sport utility vehicle [3]. Fifth, the genetic algorithm search parameters utilized in this application example are a population size of 150, 600 generations, a mutation rate of 0.01, two-point crossover, and a crossover rate of 0.75. These parameters were selected by analyzing the optimization performance of a large number of experiments that evaluated a wide range of the aforementioned optimization parameters.

The number of decision variables and their feasible input values results in a large search space of 21.4 million possible solutions, where each solution represents a unique combination of site layout and security decisions. The optimization model was utilized to search this large and highly constrained solution space in order to identify near-optimal solutions. Accordingly, the model generated a wide spectrum of 45 Pareto-optimal (i.e. nondominated) solutions, where each solution represents a unique and optimal tradeoff between the two optimization objectives of minimizing total site destruction and minimizing site construction cost, as shown in Fig. 2. The generated pareto-optimal solutions range from a minimum destruction level of 8.11% at a maximum construction cost of $6.50 million (solution S1) to a minimum construction cost of $3.96 million with an expected destruction level of 53.9% (solution S45).

Table 1. Sample building input data.

ID	Description	Dimensions		Importance weight (w_i)
		Length (m)	Width (m)	
F1	Dormitory	15	10	0.10
F2	Dormitory	15	10	0.10
F3	Dormitory	15	10	0.10
F4	Dormitory	15	10	0.10
F5	Kitchen and dining facility	12	9	0.05
F6	Communications building	8	8	0.15
F7	Headquarters	12	6	0.15
F8	Senior leader dormitory	15	10	0.10
F9	Storage	8	8	0.01
F10	Maintenance	8	8	0.04
F11	Emergency response	12	9	0.1

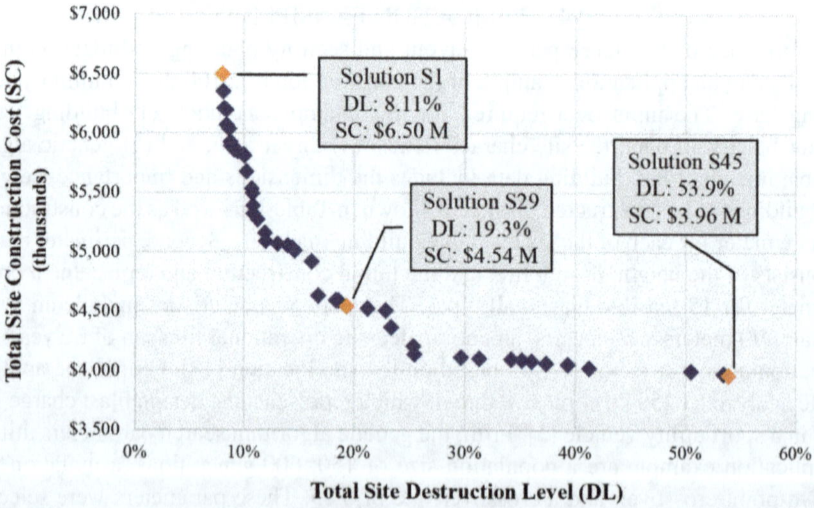

Figure 2: Generated Pareto-optimal solutions.

The minimal destruction level solutions (S1) is achieved by selecting the perimeter blast wall type and building materials that provide the greatest levels of blast mitigation and resistance: a five-meter tall, thick sand blast wall and buildings constructed of steel frame with lightly reinforced CMU infill walls or reinforced concrete with lightly reinforced CMU infill walls, as shown in Fig. 3. At the other end of the spectrum, the minimal cost solution is achieved by constructing the most inexpensive perimeter blast wall type considering the site's 30-year lifespan, a two-meter-tall, thin precast concrete panel wall and utilizing the most inexpensive building materials, wood-frame and pre-engineered metal construction. Between these two ends of the spectrum, the model generated 43 other nondominated solutions that

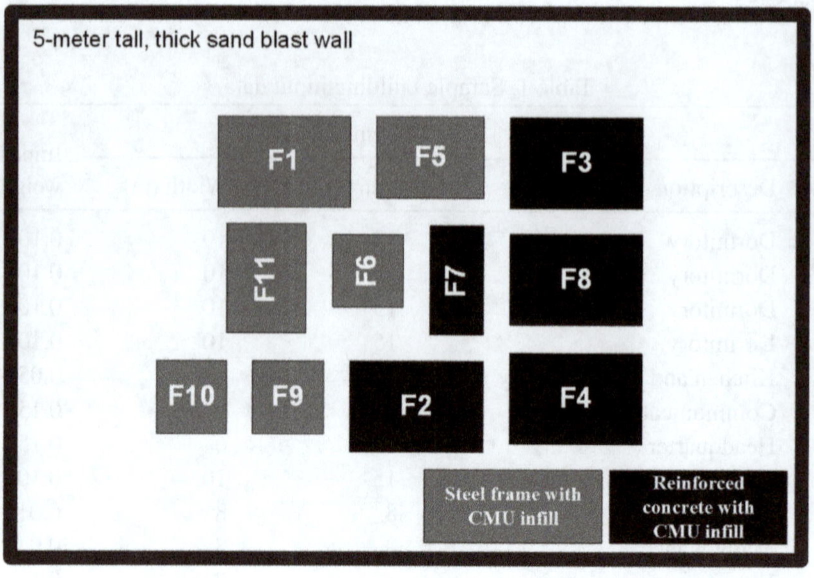

Figure 3: Site layout plan with the lowest total site destruction level (S1).

designers can select based on their maximum acceptable site destruction level in the event of an explosive attack and their maximum available construction budget. For example, the design that results in the lowest possible construction cost ($4.54 million) where the maximum acceptable destruction level is set to 20% is identified as solution S29 in Fig. 2.

5 SUMMARY AND CONCLUSIONS

This paper presented the development of a new multi-objective optimization model capable of optimizing the site layout and security planning of critical building and infrastructure systems in order to simultaneously reduce site destruction levels in the event of an explosive terrorist attack and reduce site construction costs. An application example was evaluated in order to highlight the significance and usefulness of the developed model. The model was able to efficiently search a large solution space and identify 45 near-optimal site layout and security plan solutions for multiple explosive threats. This capability enables designers to analyze and select the optimal design solution that best meets the mission requirements of their site based on the anticipated explosive threat and their available construction budget. The scope of the developed model can be expanded in future research studies to consider and evaluate the consequences of explosive attacks inflicted on site personnel and operations.

ACKNOWLEDGEMENTS

The authors would like to thank Dr. Ahmet Soylemezoglu and the rest of the virtual Forward Operating Base team from the U.S. Army Engineer Research and Development Center, Construction Engineering Research Laboratory for their insightful comments and recommendations. The views expressed in this paper are those of the authors and do not reflect the official policy or position of the United States Air Force, Department of Defense, or the United States Government.

REFERENCES

[1] Hunter, R.J., *Perkins, explosive states: AOAV's Explosive Violence Monitor 2014*, AOAV, 2015, available at: https://aoav.org.uk/2014/aoav-explosive-violence-data-2014/. (accessed January 28, 2016).

[2] Institute for Economics and Peace, *2015 Global terrorism impact: measuring and understanding the impact of terrorism*, IEP, 2015, available at: http://economicsandpeace.org/wp-content/uploads/2015/11/2015-Global-Terrorism-Index-Report.pdf. (accessed December 12, 2015).

[3] Federal Emergency Management Agency (FEMA), *Reference manual to mitigate potential terrorist attacks against buildings, FEMA-426BIPS-06 Ed. 2.* Washington, DC, 2011.

[4] Department of Defense (DoD), *DoD minimum antiterrorism standards for buildings*, UFC 4-010-01. Washington, DC, 2012.

[5] Rose, T.A., Smith, P.D. & Mays, G.C., The effectiveness of walls designed for the protection of structures against airblast from high explosives. *Proceedings of the Institution of Civil Engineers - Structures and Buildings*, **110**, pp. 78–85, 1995. https://doi.org/10.1680/istbu.1995.27306

[6] Rose, T.A., Smith, P.D. & Mays, G.C., Protection of structures against airburst using barriers of limited robustness. *Proceedings of the Institution of Civil Engineers - Structures and Buildings*, **128**, pp. 167–176, 1998. http://dx.doi.org/10.1680/istbu.1998.30123

[7] Beyer, M.E., *Blast loads behind vertical walls*, DTIC Document, 1986, available at: http://oai.dtic.mil/oai/oai?verb=getRecord&metadataPrefix=html&identifier=ADP005331 (accessed January 29, 2016).

[8] Scherbatiuk, K. & Rattanawangcharoen, N., Experimental testing and numerical modeling of soil-filled concertainer walls. *Engineering Structures*, **30**, pp. 3545–3554, 2008. https://doi.org/10.1016/j.engstruct.2008.05.030

[9] Bogosian, D. & Piepenburg, D., Effectiveness of frangible barriers for blast shielding. In *Proc. 17th Int. Symp. Mil. Asp.* Blast Shock, Las Vegas, Nevada, USA, 2002.

[10] Chen, L., Zhang, L., Fang, Q. & Mao, Y., Performance based investigation on the construction of anti-blast water wall. *International Journal of Impact Engineering*, **81**, pp. 17–33, 2015.
https://doi.org/10.1016/j.ijimpeng.2015.03.003

[11] Khalafallah, A. & El-Rayes, K., Minimizing construction-related security risks during airport expansion projects. *Journal of Construction Engineering and Management*, **134**, pp. 40–48, 2008.
http://dx.doi.org/10.1061/(ASCE)0733-9364(2008)134:1(40)

[12] Said, H. & El-Rayes, K., Optimizing the planning of construction site security for critical infrastructure projects. *Automation in Construction*, **19**, pp. 221–234, 2010. https://doi.org/10.1016/j.autcon.2009.10.005

[13] Li, Z., Shen, W., Xu, J. & Lev, B., Bilevel and multi-objective dynamic construction site layout and security planning. *Automation in Construction*, **57**, pp. 1–16, 2015. https://doi.org/10.1016/j.autcon.2015.04.011

[14] Deb, K., Pratap, A., Agarwal, S. & Meyarivan, T., A fast and elitist multiobjective genetic algorithm: NSGA-II. *Evolutionary Computation*, **6**, pp. 182–197, 2002. https://doi.org/10.1109/4235.996017

[15] Schuldt, S. & El-Rayes, K., Quantifying blast effects on constructed facilities behind blast walls. *Journal of Performance of Constructed Facilities*, **31**(4), 2017.
http://dx.doi.org/10.1061/(ASCE)CF.1943-5509.0001015

[16] U.S. Army Corps of Engineers (USACE), *Estimating damage to structures from terrorist bombs field operations guide, ETL 1110-3-495.* Washington, DC, 1999.

[17] Khalafallah, A.M., Assessing the performance of the non-dominated sorting genetic algorithm in optimizing construction site planning. *Computing in Civil and Building Engineering*, **2014**, pp. 1254–1261, 2014.
http://dx.doi.org/10.1061/9780784413616.156

[18] U.S. Army Corps of Engineers (USACE), *DoD Area Cost Factor*, 2016, available at: http://www.usace.army.mil/Cost-Engineering/Programming-Administration-and-Execution-System-Ne/. (accessed February 21, 2017).

PREVENTIVE PLANNING MODEL FOR RESCUE PRIORITY MANAGEMENT IN SEISMIC EMERGENCY

MONICA CARDARILLI, MARA LOMBARDI & MASSIMO GUARASCIO
Department of Chemical Materials Environmental Engineering, Sapienza University of Rome, Italy.

ABSTRACT

This study stems from the detection analysis of damaged buildings in Nocera Umbra (Central Italy), which has been affected by the earthquake sequence during 1997–1998, causing serious damages to material goods and human casualties. This natural disaster highlighted the need to achieve a strategic planning during the emergency phase based not only on effective actions but also on the efficient management of rescue and relief operations.

The application of Quantitative Risk Analysis (QRA) has led to the seismic risk assessment for estimating the potential level of the expected damage, on heritage buildings and human lives.

Data and information have been gathered during the surveys carried out after the major seismic event. A deep analysis has highlighted the existence of a correlation ratio between the critical structural features and the propensity of buildings to collapse. Considering either the exposed population, this information represents the crucial element for a relief management plan.

All risk indicators associated to each damage scenario have allowed to assign a scale of priorities at every relief operations for an effective risk mitigation based on damage level and expected casualties.

The ex-post emergency organization would be able to make more efficient both allocation and management of resources, equipment and technical teams during rescue activities. This is strongly connected to the ex-ante priority planning of the operations related to magnitude and distribution of expected collapses and casualties thank to the simulation performed to the potential risk scenarios.

Keywords: building collapse, Quantitative Risk Analysis, seismic emergency, vulnerability index

1 INTRODUCTION

Italy is a Country with high seismicity, due to the frequency and strength of the earthquakes that have affected its territory and we know the effects thanks to the written testimonies left by many eyewitnesses over the centuries that have allowed us to reconstruct the earthquake [1]. Being a Country subjects to earthquake, quantitative risk measurement is an imperative necessity for optimum emergency management and for the controlled planning of costs for the prevention and protection of individuals and cities [2]. The earthquake manifests itself as a rapid and violent ground shake and occurs unexpectedly without notice. It is very hard, but its consequences, in terms of victims, material damage and population involved, are often dramatic.

An earthquake is characterized by a sequence of shocks called seismic period, which sometimes precedes and almost always follows the main shock. The oscillations caused by the passage of seismic waves cause horizontal impetus on the constructions, causing serious damage or even collapse if they are not constructed with anti-seismic criteria.

In addition, the shaking of buildings is not always the same, but it depends on the local conditions of the area, in particular the type of land on the surface and the shape of the landscape [3].

For the evaluation and management of territorial risk due to the earthquake, the methodology of Quantitative Risk Analysis (QRA) has been applied which enables an integrated multidisciplinary analysis of the factors that contribute to the assessment of the risk in the territory of the municipality of Nocera Umbra. Seismic risk, so that it may be addressed and managed at best, needs to be studied and analyzed through contributions from a variety of

DOI: 10.2495/SAFE-V8-N2-307-319

disciplines and sectors such as seismology, geophysics, geomatics, structural engineering, and emergency.

2 GEOLOGICAL SETTING AND SEISMIC ACTIVITY

Nocera Umbra is a municipality in the province of Perugia. The town is located in the centre of a vast mountainous area, along Via Consolare Flaminia, which connects the towns of Foligno and Gualdo Tadino. Located on the top of a rock spur, dominated by the small River Topino and the Flaminia, Nocera Umbra, with its perched appearance, shows clear signs of its past defensive structure.

Nocera Umbra is characterized by the expanse of lithological formations ranging from Red Scaglia to Marnoso-Arenacea, attributed to the Appennines series.

The latter shows the alternation of sedimentary rocks (limestone, marl, clayey and marble limestone), each with different lithological and mechanical characteristics, above which there is a layer of heterogeneous soil with poor mechanical properties. There are therefore differently erodible soils.

Seismic Microzonation evaluates site local effects as those effects that involve changes in the seismic wave propagation (amplification or reduction) in the more superficial layers of soil. Seismic Microzonation studies allow to characterize the territory by identifying and delimiting areas with homogeneous behaviour, distinguishing between stable zones, stable zones Susceptible to local amplification and areas subject to instability. For Nocera Umbra, the amplification zones have been obtained using the map of the amplification zones, as illustrated below.

In the present study, we will use the FA amplification factors from the detailed microsynculation investigations carried out [4].

Started in September 1997, the earthquake struck the fascination of the Apennine chain between Marche and Umbria, already home to destructive earthquakes.

The intensity of the strongest shock (5.8 Richter - IX Mercalli) is in line with those of the historical earthquakes that have affected East Umbria in the past.

3 SEISMIC RISK ASSESSMENT

Italy is one of the most seismic countries in the Mediterranean because of the frequency of earthquakes that have historically affected its territory and the intensity that some of them have achieved, resulting in a significant social and economic impact.

The seismic risk assessment for Nocera Umbra was conducted using the QRA methodology, in particular by using the ETA (Event Tree Analysis).

This analysis, in fact, is based on a logical-intuitive structure. In addition, since the succession of steps describing the evolution of the danger flow is chronological, it is easy to traverse the chain of events that derive from an initiating event and reproduce the flow itself. It is in fact, as a sequential and interconnected representation of all possible alternative paths of evolution of danger scenarios triggered by a seismic event. Each event consequence is associated with a conditional probability from all previous events in the branch [5].

3.1 Initiator event probability

The knowledge of the Italian seismicity is made possible by the great number of studies and documents about the effects that earthquakes have provoked in the past in the various geographical areas. Especially in recent years, these studies have been used in regional analysis for zoning (seismic classification) or Microzonation. In the latter case, the assessment of the hazard involves the identification of areas which may, during a seismic shock, be subject to amplification phenomena.

The likelihood of occurrence associated with the initiating event may be evaluated from the probabilistic maps of the seismic hazard of the national territory, expressed in different shaking parameters on a square mesh grid of approximately 5 km per side, provided by the INGV (National Institute of Geophysics and Vulcanology). The maps used for the present work report the maximum horizontal acceleration of the ground as defined by [6] as the shake parameter, corresponding to PGA.

These maps have been calculated for different probability of overcoming in 50 years. For each estimate, we get the 50th percentile distribution (median map, which is the reference map for any probability of overcoming). Thus, Hazard Curves have been reconstructed showing the PGA trend for the various annual overcoming probabilities needed to determine the likelihood of occurrence of the Initiator Event (P(EI)).

By constructing a semi-logarithmic graph that has ascending times and ordered PGA thresholds, it is possible to determine the return time, starting with the PGA value of the event of September 26th, 1997 for Nocera Umbra, obtained using the Attenuation Law [7], which was derived from a European database consisting of 422 recordings generated by 157 earthquakes measuring magnitude Ms between 4.0 and 7.9, and a focal depth of less than 30 km.

$$\log(PGA) = -1.48 + 0.266M - 0.922logr + 0.117S_A + 0.124S_S \pm 0.25$$

where:

M = registered Magnitude

r = $(d^2 + 3.5^2)^{0.5}$

d = focal depth (i.e. the distance between the Epicenter and the Hypocenter)

In the semi-logarithmic chart below, the ascending times are reported in relation to the different probabilities and the PGA thresholds for the 50th percentile are ordered.

From the intersection with the interpolation line, the corresponding return time was obtained (See Fig. 1).

Earthquake interception times follow the statistical distribution of a Poisson process. This means that each seismic event in an area is independent in time and space from previous events in the same area and does not affect spatial location, time of occurrence, and size

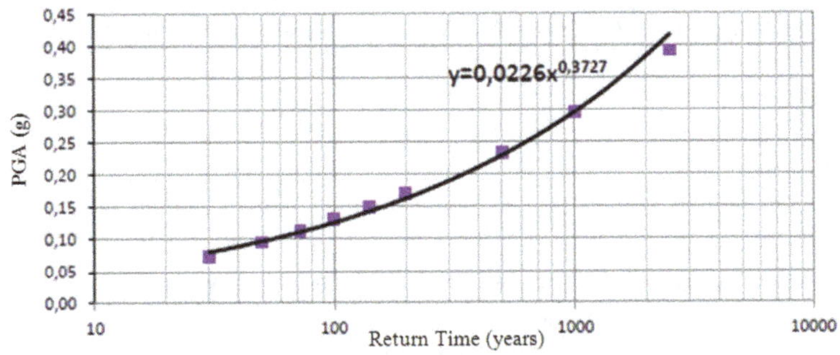

Figure 1: 50th percentile curve.

(Magnitude) of subsequent ones. Thus, the probability of the Initiator Event is given by the following equation:

$$\mathrm{P}\left(EI\right)=\frac{(\lambda t)^{n}}{n!}e^{-\lambda t}=1.06*10^{-2}\left[\frac{events}{year}\right] \quad (2)$$

Initiator Events are the starting point from which the Event Tree is drawn and represent the events that precede or start the chains of subsequent events. Starting from them, we identify the different scenarios of damage that depend on the severity of the consequences.

3.2 Propensity to collapse

The first Tree node represents the probability of the collapse event occurring for a given building. This is affected by several variables, the structural vulnerability of the building itself and the amplification of seismic shock, determined by the morphology and geological characteristics of the soil.

The estimation of structural vulnerability is a necessary step to understand the response of a building to seismic stress, that is, its propensity to damage in case of an earthquake [8].

The sample of examined buildings consists of 150 buildings, all of them masonry, located in Nocera Umbra. The choice of the sample was dictated by the desire to treat together buildings of the same type of construction and the fact that masonry buildings, both ordinary and monumental historical value, are one of the most vulnerable artefacts of the earthquake [9].

For the calculation of the vulnerability of the sample examined, the data in the 1st level detection, damage and promptness data for ordinary buildings in post-seismic emergency (AeDES Forms) were used. In activities that characterize the post-earthquake, a significant moment is usually represented by damage relief and the assessment of seismic agility. Post-seismic emergency agility assessment is a temporary and expeditious evaluation formulated on the basis of an expert judgment and conducted in a limited time. It is based on simple visual analysis and the collection of easily accessible information to determine whether buildings affected by the earthquake may be used while reasonably protected human life. AeDES Forms therefore quantify the damage reported by each building on the basis of the outcome of the assessment.

3.2.1 Building Features and Numeric Encoding

In order to arrive at the determination of the probability of collapse for each of the buildings under study, several different characteristic factors must be considered for each of them. The parameters assumed at the base of the vulnerability study were identified as more important factors for the description of the seismic behaviour of masonry buildings:

- Number of floors;
- Regularity of shape;
- Period of construction;
- Position in the aggregate.

The first two give information about building form and size, while the second ones about resistance and stability characteristics.

The height of the building has a significant influence on the resistance of the structures under the action of seismic motions. In fact, it plays a significant role under the effect of seismic stresses [10].

Regularity in plan and height is an essential requirement for defining the seismic response of a building. An uniform distribution of strength and stiffness along the height of the building is a significant factor for the good seismic behaviour of the structure. Regularity in fact avoids the development of asymmetries and hence the formation of plastic hinges (weak points) resulting in an uneven concentration of agent stresses [11].

Construction period parameter takes into account existing regulations at the time of construction and therefore the construction techniques and the anti-seismic technologies used. For this reason, it was considered appropriate to take into account the main technical legislations for masonry buildings [12, 13].

Finally, it is also important to identify the location of the building within the structural aggregate: isolated, internal, at corner, at edge for considering possible interactions from structural contiguity between adjacent buildings (hammering phenomena).

Each property of the building has been divided into three classes and each one has been assigned to a category (red, yellow, green). The categories were attributed to the contribution of the individual class to the overall structural integrity of the building.

Using the values of post-earthquake damage reported on the AeDES Forms, a numerical conversion was performed by assigning to each of them a coding '0' no collapse, and '1' collapse, in order to achieve a dichotomous classification of the damage reported by the buildings, as below:

- Encoding '0' – building does not collapse:
  - null;
  - slight damage;
  - average damage.
- Encoding '1' – building collapses:
  - serious damage;
  - very serious damages;
  - destroyed.

The correlation coefficients for the four parameters, previously standardized, were calculated. A correlation analysis was carried out between all the variables considered.

The variance-covariance matrix was calculated, which coincides with the matrix of correlation in our case. It is possible thus to evaluate how much a pair of parameters are correlated, and the sign indicates whether the correlation is positive or negative.

The major correlation is between the position and the construction era, which are the parameters that provide information on the building's resistance. In particular, it turns out that the most recent buildings are in a corner or end position in the aggregate.

A good correlation is also between the building age and the number of plans: that is what it is indicating that most recent buildings have more floors.

Table 1: Correlation matrix.

	Position	n. Floors	Age	Regularity
Position	1	0.016	0.431	0.101
n. Floors	0.016	1	0.164	0.028
Age	0.431	0.164	1	0.024
Regularity	0.101	0.028	0.024	1

One further consideration that emerges is that recent buildings are also the most irregular.

To determine the weights to associate with each parameter, two different approaches have been used:

1. Principal Components Analysis (PCA);
2. Normalized-Correlation with 'Collapse - does Not Collapse (C/NC)' variable.

3.2.2 Correlation analysis of collapsing

From the variance-covariance matrix, which coincides with the correlation matrix in the case of standardized variables, the main self-extractor is extracted to highlight the covariates between the variables considered. This was obtained by the factorial analysis method performed with the data processing software R.

The main driver co-ordinates represent the weights of the coded parameters. The positive weight indicates that the coarse reference parameter in the same sense with the other three, that is, the combined effect of the interactions and covariation is concordant. Another weighing system is the normalized correlation coefficients of each single parameter with the standard collapse/no-collapse variable.

The obtained variable was correlated with the four characteristic parameters of the buildings.

The parameters were taken individually to study how each of them influences propensity to collapse. From the weight analysis of the correlation coefficients, it may be seen first that they are all differentiated between them and with not very high values.

The position is the most linked feature of the propensity to collapse, while the number of floors is the least-related parameter. This is justified by the fact that the buildings, being in masonry, do not have a large variation of the floors parameter: most of it consists of two, three or four floors. So, as far as brick buildings are concerned, the collapse will be associated with other variables.

In order to have a better visualization of the weights thus obtained, the percentages of buildings for the individual parameter encodings for the collapse/no-collapse variable have been calculated. The relative percentages of collapsed buildings were also evaluated.

3.2.3 Evaluation of the Vulnerability Index

Thus, with both the methodologies, the Vulnerability Index (Iv) for each building was calculated by adding the products between the weights obtained and the standardized encoded values of the building parameters. It is denoted that both calculated Iv have a percentage of 50% explained. Since the apparent capability of an Iv range has not been evident from the results of the calculated vulnerability indices, the graph has been comparing the two Indices thus obtained and returning the number of coincident values for each identified coordinate.

The value of the Iv, valued by correlation coefficients, is higher than the equivalence condition represented by the other one.

3.2.4 Regression model

Through the use of regression models, it was possible to extrapolate the value of the propensity to collapse, starting with the calculation of the vulnerability, using the data reported on the damage relief cards. To get the propensity to collapse from the estimated values of vulnerability and the collapse/no-collapse variable, the regression models have been applied.

The application of regression models to buildings surveyed with damage relief forms allows to estimate the probability of collapse depending on the vulnerability values and site amplification.

To analyze a dichotomous variable, we use Logistic regression, which represents a Generalized Linear Model (GLM) in which the link function is the 'logit' of the probability of belonging to a category (code '1') with respect to not belonging to it (code '0').

The propensity to collapse is then defined by the following equation:

$$\theta = \mathrm{P}(Y = 1) = \frac{\mathrm{e}^{(\beta_o + \beta_1 x_1 + \beta_n x_n)}}{1 + \mathrm{e}^{(\beta_o + \beta_1 x_1 + \beta_n x_n)}} \tag{3}$$

where x_i represents the building vulnerability and local amplification, and β the regression coefficients.

The statistical significance (p-value) of the regression coefficients obtained from the first processing is higher than the respective value reported by the regression of collapse/no-collapse variable where both fall within the limit value of 0.05.

The second elaboration of the vulnerability estimated by correlation coefficients shows a significant increase in the statistical significance of the variable, hence the ability of the variable to explain the propensity to collapse.

The AIC index allows to evaluate the adaptation of data with the link used. The AIC indicator value from both outputs indicates a discrete adaption of the logit model to the data. The AIC of the second implementation also shows an improvement in model adaptation to data compared to the previous case.

The Iv model provides a p-value and a lower AIC, hence a greater statistical significance of the performed estimate and better model quality than the data implemented.

It may be observed that the highest values that characterize the range of the estimated vulnerability with the correlation coefficients are also reflected in its propensity to collapse. In particular, the range of the right table is also characterized by greater explanatory capacity, with a level of explanation of 61% versus 44% for the two different implemented Iv.

Since the Factor of Amplification plays an important effect, this is reflected in the value of the collapse probability, in fact the propensity values are increased in the minimum and maximum value of the range.

Based on the results of regression, it is considered that the most effective Iv in explaining collapses is the one resulting from second processing, as expected.

Regression models therefore allow for an estimate of the propensity to collapse that takes into account local seismic microzonation by providing an assessment of the most representative buildings and hence a more effective preventive action and/or more aware emergency management in the post-earthquake.

Table 2: Propension to collapse

	Vulnerability		with Amplification
	PCA	C/NC	C/NC
min	0.07	0.08	0.33
max	0.51	0.69	0.93

3.3 Salvability estimation

In the second and third bifurcations, crowding and exposure are considered. The second Tree node represents the exposure, expressed in this specific case in terms of human lives, as for civil protection purposes it is necessary to estimate the social losses expected as a result of an earthquake [14].

The starting data required for the quantification of the exposed to a given seismic event is that of the population resident in the analyzed buildings [15]. The population per building varies considerably depending on the location and is a function of the characteristics and dimensions of the buildings. To determine this parameter may be used the data provided by the ISTAT Census or the data reported on the AeDES Forms where available, as in our case. From the AeDES Forms, the number of occupants for each building was estimated by the exhibitors.

The earthquake event timetable is a determining factor for the estimation of the victims, as it affects the number of people present inside the buildings at the time of collapse [16].

In the construction of Events Tree must be considered two scenarios: daily and nightly by using Coburn model [17] applied to residential buildings.

The mediator most likely to influence the likelihood of an exposed individual being involved in the collapse is represented by the building and its vulnerability. In fact, injuries caused by the involved are caused by the size and mode of collapse of the building, depending on the type of structure, intensity and type of construction shock [18].

In the Coburn model, the parameters to be considered for determining the number of victims among the Exposed (E) people, in the case of masonry structures such as those in the present study, are as follows:

- Trapped (I): 60% * E
- Victims (V): 20% * I

By applying these reductive factors, the number of deaths associated with the earthquake is obtained according to the scenario assumed, which is indispensable for the quantification of the damage, both expressed in terms of human lives that in economic terms associated with one or more seismic events analyzed.

The second mediator affecting the salvability of these individuals is the management of relief, in fact, the lengthening of survival research times greatly exacerbates the likelihood of getting them out of life again [19].

In the construction of the Event Tree, a parameter is used that takes into account the speed of the intervention of the rescuers, assuming that it takes place within the first 48 h of the earthquake, beyond which the probability of survival is reduced to values below 30%.

3.4 Risk indicators and acceptability thresholds

In the last bifurcation is considered the efficiency indicator of emergency management that takes into account the rescue and self-relieving (salvability).

From Event Tree Analysis, it is possible to determine the probabilities of all possible end-of-turn scenarios [20]. From the knowledge of the latter, the expected value for the scenario considered was the number of expected casualties [21].

Afterwards, the retrospective distribution was determined, which provides information on the social perception of risk (Social Risk).

Social risk perception may be expressed in terms of the Social Damage Curve, where the probability of overcoming the corresponding threshold in the order of P is associated with each possible degree of damage expressing the number of expected casualties N.

In order to develop prevention and mitigation policies, it was considered necessary to make the assessment scenario more complete by comparing the expected Risk Indicator with an appropriate limit value of the trend of the retrofitted distribution by means of curves representing tolerance limits considered socially acceptable or not acceptable.

The red line represents the threshold of unacceptability, while the green line is the threshold of acceptability.

Analyzing the graph above, we may see that the distribution of the retrofitting is completely within the threshold, although its performance is close to the straight line unacceptability.

It should be noted that the levels of acceptability of the cumulative risk used in the present study are those related to railway tunnels [22], which provide maximum values of acceptability and unacceptability of probabilities equal to 10^{-4} and 10^{-2}.

The area underneath this curve represents the Expected Value E(N), the Risk Indicator as sum of the number of casualties expected for the respective scenario.

From the graphic representation below, we may see the probability of achievement and/or overcoming the expected number of victims for a given PGA value (See Fig. 2).

3.5 Mosaic model

For a rational and effective emergency management in a seismic event, it is necessary first to subdivide the whole urban grid into sub-areas, in order to associate one or more of them with each rescue team through the application of the 'Mosaic Model'. The territory on which buildings fall into is to be divided into 'tiles', or irregular polygons delimited each by urban road. The tiles at first are defined according to the roads deemed feasible in the event of an emergency.

In the specific case, having analyzed a small number of buildings, there were areas of intervention containing inside a comparable number of buildings.

The tiles, uniquely defined with increasing numbering, are the basic element on which to carry out estimates of the quantification of the victims and the injured in order to order the tiles on the intervention scale and associate them with the rescue teams [23]. Sending human resources and equipment is a priority in those cards that present more trapped and injured.

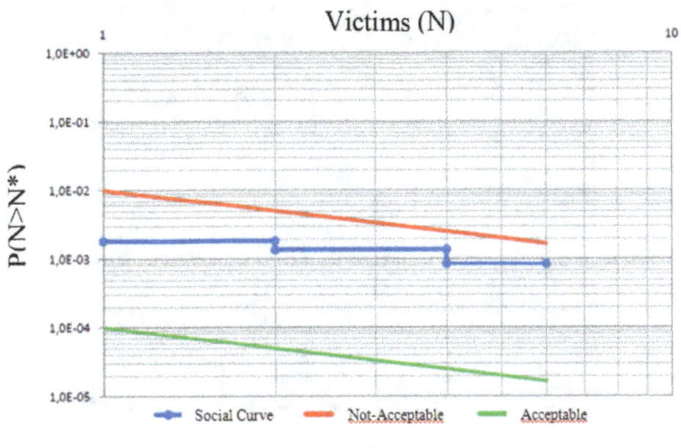

Figure 2: Social damage curve with acceptability thresholds (for a type building).

Adding the wounded from the Event Tree for the buildings belonging to the same tile, the number of injuries was determined for each of them. In this way, it was possible to determine the intervention order of the rescue teams, starting from the tile with the most injured to the one with the least number (See Fig. 3). Tiles play a key role in organizing relief, estimating the number of injuries expected and sending the number of intervention teams needed. With the QGIS software, the tiles and their associated buildings were represented.

Other information that may be given to the tiles in addition to the number of wounded is the probability of collapse associated with each building for a previously assessed PGA date. This would provide useful guidance to rescue teams already in the early hours of the event, during which the rescuers work uninterruptedly to help the trapped potentially.

The evaluations were carried out for the most critical tile since it is characterized by buildings with maximum and minimum probability values, for the case study, and because it is ranked first in the ranking for the intervention order and therefore by the number of inhabitants.

The maximum and minimum trend of the tile has also been analyzed as a probability of overcoming the back-cumulated curve (See Fig. 4). In this case, the scenario of damage is represented by the collapse of buildings. The damage variable is between 0 (no building collapses) and 10 (collapses all buildings).

Figure 3: Intervention order for each tile.

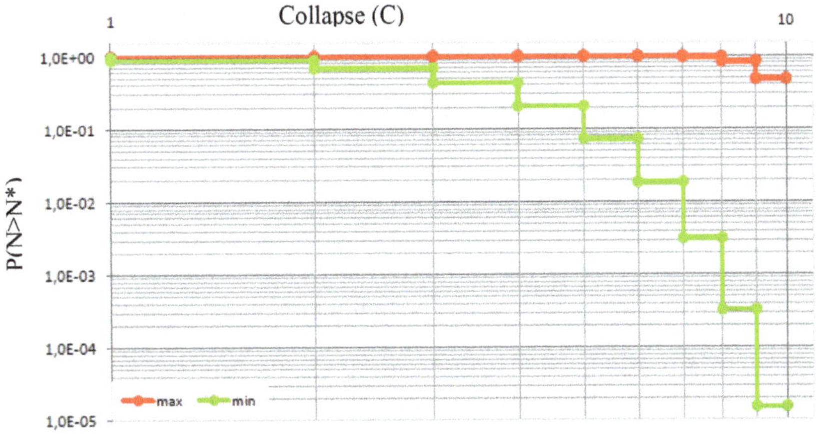

Figure 4: Maximum and minimum Social Damage Curves of the most critical tile.

4 CONCLUSION

The work involved the proposal of a methodology useful for the prevention, planning and management of the seismic emergency for the territory of the Municipality of Nocera Umbra. The analysis was carried out according to Events Tree that allows to determine collective and individual Risk Indicators based on the expected number of victims. Risk indicators have been evaluated either by referring to the victims or by the collapse of each mosaic 'tile' in the city itself: in the first case, the areas at greater risk within the urban area; in the second allows to compare the level of risk of different buildings.

The indicators would then allow for any intervention priorities for relief to reduce seismic risk. Inside each tile, in fact, the buildings that could have the largest number of victims were identified. It will be exactly those in which the team will intervene at the first moment of the emergency.

The data thus obtained, collected in the form of databases, represent an indispensable tool for timely and effective seismic emergency management.

The application of the described model and the 'mosaic' approach to an entire urban nucleus would allow for a planning of the emergency that is capable of first dimensionizing the resources needed to manage the post-earthquake at the variation of the magnitude of the shock.

Anyway for a more comprehensive and exhaustive analysis, if good quality data is available, other parameters such as degree of apprehension, height interception, site morphology etc. should be considered.

REFERENCES

[1] Ariki, F., Shima, S. & Midorikawa, S., Earthquake disaster prevention of Yokohama City. *Journal of Japan Association for Earthquake Engineering*, **4**(3), pp. 148–153, 2004.
https://doi.org/10.5610/jaee.4.3_148

[2] Bal, I.E., Crowley, H. & Pinho, R., Displacement-based earthquake loss assessment for an earthquake scenario in Istanbul. *Journal of Earthquake Engineering*, **11**(2), pp. 12–22, 2008.
https://doi.org/10.1080/13632460802013388

[3] Erdik, M., Sesetyan, K., Demircioglu, M.B., Hancılar, U. & Zulfikar, C., Rapid earthquake loss assessment after damaging earthquakes. *Soil Dynamics and Earthquake Engineering*, **31**(2), pp. 247–266, 2011.
https://doi.org/10.1016/j.soildyn.2010.03.009

[4] GNDT/SNN Ratio Seismic Microzonation Project Umbria-Marche Nocera Umbra: Seismic Microzonation Elements, Milan, 1999.

[5] Spence, R.J.S., Coburn, A.W., Sakai, S. & Pomonis, A., A parameterless scale of seismic intensity for use in seismic risk analysis and vulnerability assessment. In: Society for References Earthquake and Civil Engineering Dynamics (ed.), *Earthquake Blast AND Impact: Measurement and Effects of Vibration*. Elsevier Applied Science, Amsterdam, 1991.

[6] OPCM (2006) n. 3519

[7] Ambraseys, N.N., Simpson K.A. & Bommer, J.J., Prediction of horizontal response spectra in Europe. *Earthquake Engineering & Structural Dynamics*, **25**, pp. 371–400, 1996.
https://doi.org/10.1002/(sici)1096-9845(199604)25:4<371::aid-eqe550>3.3.co;2-1

[8] Calvi, G.M., Pinho, R., Magenes, G., Bommer, J.J., Restrepo-Velez, L.F. & Crowley, H., The development of seismic vulnerability assessment methodologies over the past 30 years. *ISET Journal of Earthquake Technology*, **43**(4), pp. 75–104, 2006.

[9] Lagomarsino, S. & Giovinazzi, S., Macroseismic and mechanical models for the fragility and damage assessment of current buildings. *Bull Earthquake Engineering*, **4**, pp. 445–463, 2006.
https://doi.org/10.1007/s10518-006-9024-z

[10] Larionov, V., Frolova, N. & Ugarov, A., Approaches to fragility evaluation and their application for operative forecast of earthquake consequences. *All-Russian conference* "Risk- 2000", ed. A. Ragozin, ANKIL, Moscow, pp. 132–135, 2000.

[11] Yeh, C.H., Jean, W.Y. & Loh, C.H., Damage building assessment for earthquake loss estimation in Taiwan. In: *Proceedings of the 12th world conference on earthquake engineering*, Auckland, New Zealand, Paper No. 1500, 2000.

[12] Italian Regio Decreto (1937) n. 2105, 22nd November.

[13] Italian Ministerial Decree (1987) n. 141, 9th January.

[14] Bommer, J., Pinho, R. & Crowley, H., Using a displacement-based approach for earthquake loss estimation. *Advances in earthquake engineering for urban risk reduction*, eds. S.T. Wasti & G. Ozcebe, Springer, Dordrecht, 2006.

[15] Crowley, H., Pinho, R. & Bommer, J.J., A probabilistic displacement-based fragility assessment procedure for earthquake loss estimation. *Bulletin of Earthquake Engineering*, **2**(2), pp. 173–219, 2004.
https://doi.org/10.1080/13632460802013388

[16] Eguchi, R.T., Goltz, J.D., Seligson, H.A., Flores, P.J., Blais, N.C., Heaton, T.H. & Bortugno, E., Real-time loss estimation as an emergency response decision support system: the EPEDAT. *Earthquake Spectra*, **13**(4), pp. 815–832, 1997.
https://doi.org/10.1193/1.1585982

[17] Coburn, A. & Spence, R., *Earthquake Protection*, 2nd edn. Wiley, Chichester, 2002.

[18] Spence, R.J.S., *Earthquake disaster scenario prediction and loss modeling for urban areas: LESSLOSS report*. IUSS Press, Pavia, 2007.

[19] Di Pasquale, G., Ferlito, R., Orsini, G., Papa, F., Pizza, A.G., Van Dyck, J. & Veneziano, D., Seismic scenarios tools for emergency planning and management. In: Proceedings of the XXIX GAESC, Potsdam, Germany, 2004.

[20] Jaiswal, K. & Wald, D., An empirical model for global earthquake fatality estimation. *Earthquake Spectra,* **26**(4), pp. 1017–1037, 2010. https://doi.org/10.1193/1.3480331
[21] Samardjieva, E. & Badal, J., Estimation of the expected number of casualties caused by strong earthquakes. *Bulletin of the Seismological Society of America,* **92**(6), pp. 2310–2322, 2002. https://doi.org/10.1785/0120010112
[22] Italian M.D. Infrastructures and Transport (2005), Annex 3, Table 4.3.1
[23] Markus, M., Fiedrich, F., Leebmann, J., Schweier, C. & Steinle, E., Concept for an integrated disaster management tool. In: Proceedings of the 13th WCEE, Vancouver, BC, Canada, 2004.

PROPOSAL FOR STRUCTURAL EVALUATION OF HISTORICAL CHURCHES AT LIMA, PERU

C. CUADRA
Akita Prefectural University, Japan.

ABSTRACT

The historic centre of Lima city, the capital of Peru, was declared a World Heritage Site in 1988 by UNESCO. The high concentration of historic monuments constructed at the time of the Spanish presence and at the beginning of the Republican era, contributes to originality and historic value of the site. Churches or convents are included in this area, and according to a list of Ministry of Culture of Peru there are 50 places considered religious monuments. As a contribution to improve urban security and safety against earthquakes, in this study, a procedure to investigate the seismic vulnerability of these churches is discussed. First, a general diagnosis of problems concerning the city and its buildings is proposed. Then a survey for preliminary evaluation of the structural condition of churches structures is planned. This evaluation of the seismic vulnerability of historic churches at the historic centre of Lima represents the basic study that is necessary to initiate detailed investigation for the preservation and conservation of these historic buildings. The study intends to establish a general guideline for vulnerability evaluation of historic churches Lima Heritage Site and it could be applied to the evaluation of other historic cities of Peru.

Keywords: ambient vibration, architectural heritage, historic churches, Lima city, seismic vulnerability.

1 INTRODUCTION

A diagnosis for evaluation of seismic vulnerability of historical building located in Lima, Peru, was performed by Cuadra *et al.* in 2013 [1]. In the present research, only historical churches are included, since these structures are more representative buildings of the historic landscape of Lima centre.

The historic centre of Lima city was declared a World Heritage Site in 1988 for its originality and high concentration of historic monuments constructed at the time of the Spanish presence and at the beginning of the republican era. The architecture of the buildings corresponds in general to typical Hispano-American baroque of the 17th and 18th centuries. Lima city was founded on January 18, 1535, by Spaniard conquistador Francisco Pizarro and played a leading role in the history of the New World from 1542, when king Carlos V of Spain establish the vice royalty of Peru, until the middle of the 18th century, just before the start of independence movements [1].

Conservation of historic buildings is an important task that implies, as a first step, the evaluation of their vulnerability. Evaluation of the vulnerability of a single building could mean the estimation of mechanical characteristics by means of detailed structural analysis or in-situ measurements. When this evaluation must be performed for a group of buildings, however, simplified evaluation is required to establish their characteristics and hence to grasp the vulnerability of target buildings. For a preliminary evaluation of structural conditions of target churches, a general survey is planned to be undertaken. For this purpose, a simple methodology for estimating earthquake resistance or vulnerability of target churches is proposed.

This evaluation of the seismic vulnerability of churches located at historic centre of Lima represents a basic study that is necessary to initiate detailed investigation for the preservation and conservation of these historic churches. A method is proposed based on simple indexes to establish the potential vulnerability of buildings. Detailed observation and in-situ

DOI: 10.2495/SAFE-V8-N2-320-328

estimation of the dynamic characteristics of critical buildings is then proposed by using ambient vibration measurements. Measurements results will permit to verify the analytical models for detailed structural analysis to establish the vulnerability of specific target buildings.

Study of vibration characteristics of soil is also a critical task to identify possible cases of resonance effect that could affect or amplify vibration response of buildings. This evaluation can be done also by means of microtremor measurements on ground surface to identify the predominant period of vibration of soil.

This process of evaluation of seismic vulnerability could also serve to establish a general guideline for vulnerability evaluation of churches or in general historical buildings located at other historic cities in Peru.

2 CHURCHES AT HISTORIC CENTRE OF LIMA

Description of characteristics of Lima historic centre and historical churches are taken from reference [1]. At the time of the founding of Lima by the Spaniards conquistadores, the site was occupied by a local government called Curacazgo and the Governor, who was the local representative of the Inca emperor, was called Curaca. Only temples called Huacas and administrative constructions existed at that time, and they were replaced by Spanish-style constructions. The city was organized in a shape of a chessboard following the original Inca distribution of streets. This portion can be appreciated in the central part of Fig. 1 enclosed by a dashed-line polygon, where straight streets and square blocks can be observed. Sometimes this portion of the city alone is called the historic centre. However, the area declared by UNESCO to be a World Heritage Site is larger, as is indicated in Fig. 1 by the area enclosed by a solid-line irregular polygon.

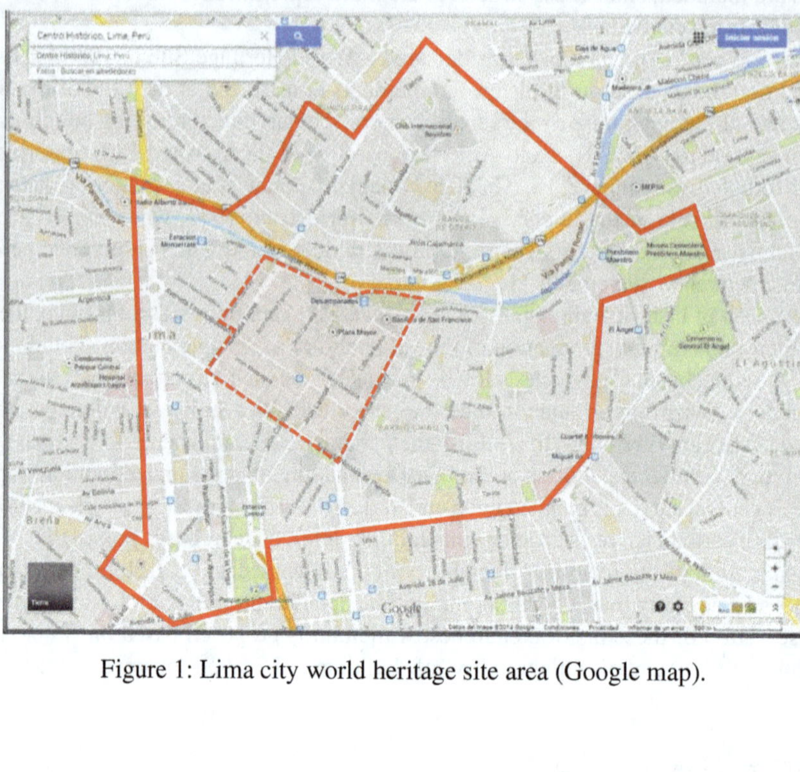

Figure 1: Lima city world heritage site area (Google map).

The Ministry of Culture of Peru has elaborated a catalogue of historic monuments located at centre of Lima. In the case of the evaluation of the vulnerability of an area of the city, characteristics of buildings located at this site must be studied, and this task is crucial in the case of historic constructions. An understanding of the characteristics of existing buildings is important to the assessment of their structural safety. At historic centre of Lima, buildings can be divided according to their use. In this case, important buildings include churches in 50 places considered religious monuments. Then there are buildings that were used as residences, institutions like banks, commercial shops, hotels, and finally some palaces or administrative offices like city hall and the presidential palace. Since there are many types of monuments, it would be adequate to choose some specific types of buildings to initiate vulnerability evaluation. Table 1 shows the places at Lima's historical centre that are religious constructions like churches, convents, and cathedrals.

Table 1: List of religious buildings at Lima's historical centre.

ID	Name	ID	Name
1	El Milagro Chapel	27	San Francisco Church
2	Santa Rosa Monastery	28	Santa Clara Church
3	T. Francisco Monastery	29	Concepcionistas Church
4	3 St. Francisco Monastery	30	Santa Ana Church
5	San Pedro Monastery	31	La Trinidad Church
6	Cathedral and El Sagrario	32	Las Nazarenas Church
7	Mercedarias Monastery	33	Las Trinitarias Church
8	San Agustin Monastery	34	Sra. El Prado Church
9	Veracruz Church	35	Santa Catalina Church
10	Santo Cristo Church	36	Santa Rosa Church
11	San Carlos Church	37	Sra. Del Carmen Church
12	Jesus Maria Church	38	Recoleta Church
13	La Merced Monastery	39	Concepcion Monastery
14	La Soledad Church	40	Bishop Palace
15	Cocharcas Church	41	Santa Rosa Sanctuary
16	Concepcion Church	42	Santo Domingo Monastery
17	Monserrate Church	43	Sra. de la Cabeza Church
18	San Agustin Church	44	Paulo Nuevo Church
19	San Marcelo Church	45	San Lazaro Church
20	San Pedro Church	46	Santa Liberta Church
21	San Pedro Nolasco Church	47	Del Rosario Church
22	San Sebastian Church	48	Sra. Copacabana Church
23	Santiago Church	49	Del Parocinio Church
24	Santo Domingo Church	50	Los Descalzos Church
25	Corazon de Jesus Church	51	San Francisco de Paula Monastery
26	La Buena Muerte Church		

In this paper, churches are proposed as target constructions because they present a critical or weak part, which is the bell tower. In the case of earthquakes, these towers failed dramatically as was observed in previous earthquakes. Figure 2 shows some churches located at Lima historic centre. Churches have one or two bell towers, which are the most vulnerable part of structure.

Lima city has suffered the effects of many earthquakes. One of them occurred in 1687 and the city was rapidly reconstructed under the influence of the baroque style that had great influence on architectural development in the next two centuries. A big earthquake that destroyed Lima and its port of Callao occurred in 1746. Callao port was also affected by a tsunami originating in the earthquake. This earthquake marks the decline in the Lima Baroque style that had been supported mainly by religious organizations. At the beginning of the republican era, new construction was influenced by New European Baroque. From 1920 to 1940, the architecture of the city was greatly influenced by English urban style based on the concept of the 'Garden City'. With the expansion of Lima, the decay of the historical centre

Figure 2: Examples of churches located at Lima historic centre.

also started, since rich and middle class persons moved to the new urban areas. The historical centre became the place of residence for the poor, who subdivided the buildings and converted them into slams. Some buildings, however, were converted or adapted to serve as commercial shopping centres and educational institutions. In the worst case these old buildings were demolished to construct apartment and office buildings, thereby distorting traditional urban space.

On the other hand, it is also necessary to consider geotechnical properties of ground foundation since soil predominant periods with similar values to predominant periods of structures can result in occurrence of resonance vibration with subsequent structural damage and even collapse. Figure 3 shows distribution of soil type of Lima city reported by Quispe [2], where alluvial gravel deposits cover large part of the city, and presents good geo-mechanic behaviour. Soil of Lima centre is characterized by relatively high predominant frequencies around 3–8 Hz. However, some measurements show also predominant peaks at about 1–2 Hz. This result suggests the presence at the surface of relatively softer soil layers that might contribute to the local response at this site, so it is required to carry out further studies in order to have a better understanding. Therefore, specific measurement of soil dynamic characteristic at each site is required.

3 PROPOSAL FOR STRUCTURAL EVALUATION

It is proposed to initiate a detailed survey evaluation to obtain general characteristics of churches. Then, based in the simplified method proposed by Lourenço and Roque [3] a first evaluation of seismic vulnerability can be done. This method considers simple indexes like In-plan area ratio, which is the relation between in-plan area of earthquake resistant walls and total in plan area of the building. The second index is the area to weight ratio, which considers the ratio between in-plan area of earthquake resistant walls and weight of the building. The last index corresponds to the base shear ratio.

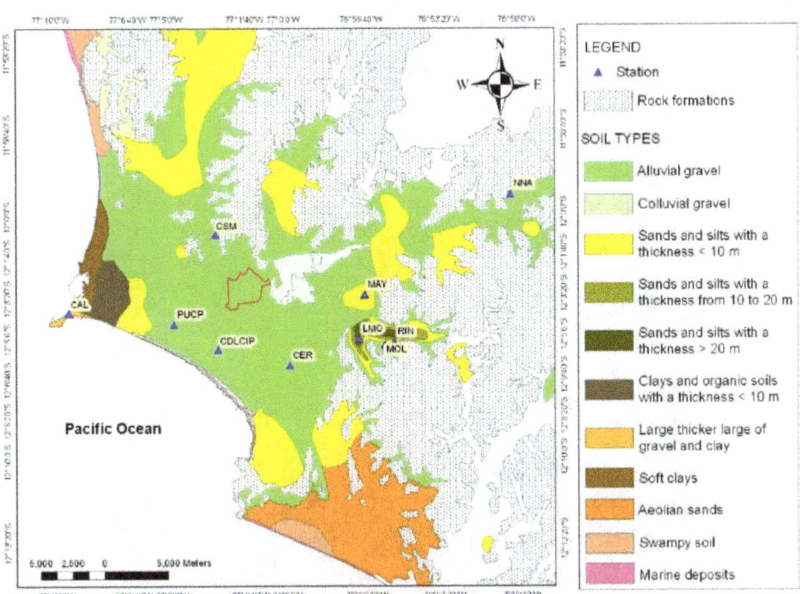

Figure 3: Type of soils at Lima city.

After the simple evaluation using simple indexes it is proposed to estimate vibration characteristics of churches using ambient vibration measurements. In this case special attention must be put on estimation of vibration of bell towers since these portions of structure are more vulnerable in case of earthquakes. Using results from ambient vibration measurements, a characterization of churches located at Lima historic centre can be done by comparing period of vibration versus structures height. In addition, characteristics of soil at each site can be obtained from ambient vibration measurement on ground.

For some specific or special churches, analysis can be done using finite element method. Reliability of model for analysis can be confirmed by comparing analytical results for period of vibration with those results obtained from ambient vibration measurements ([4–6]).

As example, results of ambient vibration measurements and finite element for Santo Domingo church located in Cusco city, Peru is presented in following paragraphs. Figure 4 shows the Church façade with the bell tower and directions of measurements are indicated by arrows.

Figure 5 presents the results of the Fourier analysis for the horizontal vibration of tower of Santo Domingo church. These directions NS and EW are indicated in Fig. 4. The Fourier amplitude spectrum obtained at the top of the tower was divided by the one obtained at the ground level, to obtain only the vibration characteristics of the structure. It can be observed that the value of the predominant frequency in the EW direction is 2.7 Hz, and in the NS

Figure 4: Santo Domingo Church, Cusco, Peru and direction of measurements.

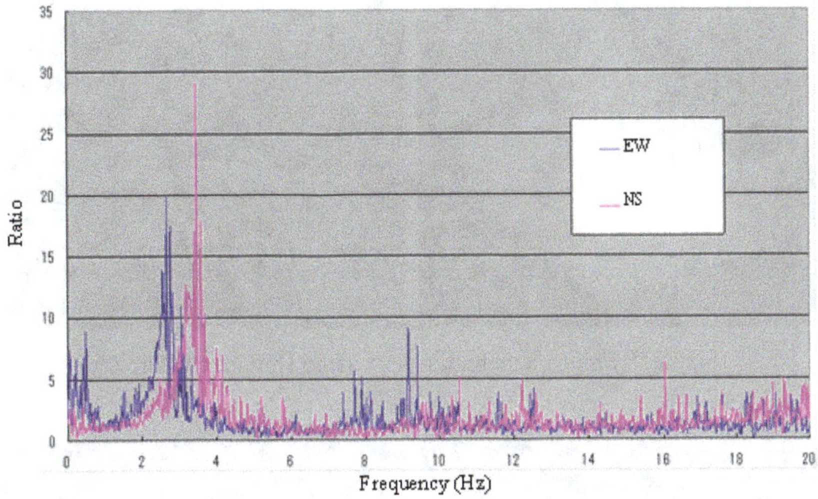

Figure 5: Fourier analysis results for the tower of Santo Domingo Church.

direction is 3.3 Hz. Figure 6 presents the dispersion curve obtained from the FK analysis. This result permits to estimate the predominant frequency of the ground as 2.7 Hz.

Comparing results of predominant frequency for ground with predominant frequencies for the tower of the Colonial Church, it can be observed that these frequencies are similar, and therefore, the resonance phenomenon is expected to occur. This coincidence of the predominant frequencies could be the explanation of the repeatedly failure of the tower during past earthquakes.

To compare the characteristics obtained from ambient vibration measurements with analytical results, finite element method is employed to formulate the analytical model. The FEM model was constructed based on available data and based on direct measurements of the dimension of the building. As can be observed in Fig. 7, the general shape of the Santo Domingo Church is reproduced in the model where only main openings (main door and windows of the tower) are considered.

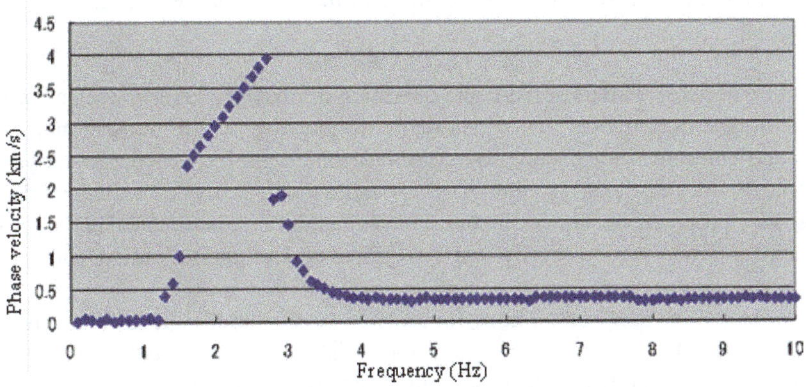

Figure 6: Rayleigh wave dispersion curve for Santo Domingo church ground site.

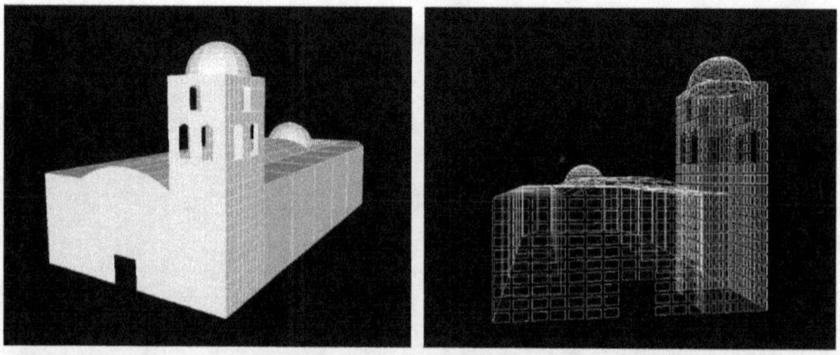

Figure 7: Finite element model of Santo Domingo Church.

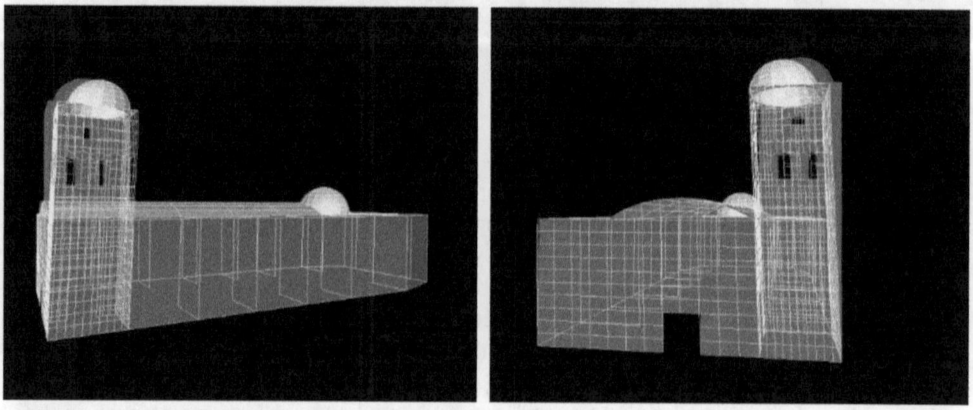

Figure 8: Modes of vibration of Santo Domingo Church.

Modal analysis using the finite element method gives as a result that the first mode of vibration corresponds to normal mode in the EW direction (or in the longitudinal direction of the church) with a frequency of 3.66 Hz. The second mode corresponds to the vibration in the transversal direction with a frequency of 3.69 Hz. These modes of vibration are show in Fig. 8 where the grey shadow indicates the un-deformed shape of the church.

4 CONCLUSIONS

A description of general characteristic and condition of historic churches located at historic centre of Lima was presented. This historical site includes 51 religious monuments like churches and convents, which have some critical conditions for structural safety. As was observed in previous earthquakes, church towers could fail in case of earthquakes. Therefore, churches were proposed as target structures for vulnerability evaluation. This proposal can contribute to improve urban security and safety against earthquakes, in special safety and security of historical centres.

As a process or research methodology for selected target churches, it has been proposed to perform first a rapid survey to establish a priority list for starting detailed evaluation. The classification of vulnerability of historic churches provides the opportunity to focus more detailed measurement on the most critical buildings. Then, for selected churches, ambient vibration measurements have been proposed to estimate dynamic characteristics of buildings, in special vibration of bell towers. From these measurements, it is possible to establish indexes of the seismic vulnerability for historic churches. It is also important to estimate vibration characteristics of soil to estimate possible cases of resonance response.

Once a list of most vulnerable churches is established, critical structures could be selected for more detailed analysis. In this case, analytical models using the finite element method or some similar method should be prepared. Ambient vibration measurement results could be used to calibrate the reliability of modelling.

This process of evaluation of seismic vulnerability could also serve to establish a general guideline for vulnerability evaluation of churches or in general historical buildings located at other historic cities in Peru.

REFERENCES

[1] Cuadra, C., Saito, T. & Zavala, C., Diagnosis for seismic vulnerability evaluation of historical buildings in Lima, Peru. *Journal of Disaster Research*, **8**(2), pp. 320–327, 2013.
https://doi.org/10.20965/jdr.2013.p0320

[2] Quispe, S., Aguilar, Z., Lazares, F., Tavera, H., Yamanaka, H. & Yamazaki, F., Evaluation of local site effects in Lima city, Peru from ground motion data, *International Symposium for CISMID 25th Anniversary: Technological advances and learned lesson from last great earthquakes and tsunamis in the world,* National University of Engineering, Lima, Peru, 2012.

[3] Lourenço, P.B. & Roque, J.A., Simplified indexes for the seismic vulnerability of ancient masonry buildings. *Construction and Building Materials,* **20**(4), pp. 200–208, 2006.
https://doi.org/10.1016/j.conbuildmat.2005.08.027

[4] Cuadra, C., Karkee, M.B., Ogawa, J. & Rojas, J., An evaluation of earthquake risk to Inca's historical constructions, *Proceedings of the 13th World Conference on Earthquake Engineering*, Vancouver, B.C., Canada, August 1–6, 2004.

[5] Cuadra, C., Zavala, C., Abe, A., Saito, T. & Sugano, S., Vibration characteristics of traditional adobe-quincha buildings located at Lima historic center, *8th International Conference on Urban Earthquake Engineering*, Tokyo Institute of Technology, Japan, 2011.

[6] Miranda, O., Gallardo, J., Cuadra, C. & Ogawa, J., Reduction of seismic damage in Peruvian traditional constructions, *2nd International Conference on Computer Simulation in Risk Analysis and Hazard Mitigation*, Risk Analysis II, Bologna, Italy, pp. 389–396, 2000.

HOSPITALS SAFE FROM DISASTERS: A GLIMPSE INTO THE MEXICAN COASTAL ZONES

MILTON MONTEJANO-CASTILLO & MILDRED MORENO-VILLANUEVA
Escuela Superior de Ingeniería y Arquitectura Unidad Tecamachalco del Instituto Politécnico Nacional, México.

ABSTRACT

Flood risks and damages due to natural phenomena are becoming more frequent in Mexican coastal zones. Given this condition, the critical infrastructure of cities must be prepared and protected for such disasters, particularly in the health sector. To face this challenge, during the World Conference on Disasters Risk Reduction in 2005, the United Nations launched the global campaign 'Hospitals Safe from Disasters'. This campaign was based on the commitment that hospitals must be safe in order to mitigate disasters damages and hospitals must continue functioning after a disaster. Mexico together with other 168 countries became part of the compromise. In 2015, said goals and objectives were ratified in the Sendai Framework for Disaster Risk Reduction 2015–2030. Under the leadership of the National System for Civil Defense Mexico´s, the 'Safe Hospital Program' has been a mechanism to evaluate and certify the safety of hospitals located on disaster risk areas. Such a program broadly speaking comprises a Hospital Safety Index, an Evaluation Process, and the corresponding Evaluators Committee. Against this background, the main goal of this paper is to show the results of a long-term project that has been documenting the limits, challenges and strategies that some hospitals pose in Mexican coastal zones in order to reach the goals of the said Program. This analysis bases on semi-structured interviews with key informants in hospitals exposed to hydrometeorological hazards as hurricanes and heavy rains. Results show that in spite of a very precise developed Safety Index and detailed evaluation tools, hospitals lack in many cases of the financial resources and a 'risk reduction culture', or have serious technical constraints imposed by the age of the buildings. Nevertheless, they have developed strategies that could be useful for other health care facilities in the same situation.
Keywords: coastal zones, disasters, hospitals, Mexico, risk reduction.

1 INTRODUCTION

On January 29, 2015, the explosion of a gas pipe inside a Mother and Children's Hospital in Mexico City caused the death of two newborns, two adults, dozens of injured and practically the complete destruction of the hospital [1]. Events like this automatically raise many questions and reflections about the safety and vulnerability in these types of buildings: Why is a Mother and Children´s Hospital so vulnerable? Why was it so difficult to quickly evacuate the hospital and especially the evacuation of the newborns? Were there no safety standards to prevent this disaster from occurring? Unfortunately this situation is not new, but the questions behind this phenomenon have been changing over time.

During the decade of 1990 (dedicated to disasters risk reduction), some disasters around the world led to a series of publications on the impacts and effects that the destruction of a hospital – by either internal or external hazards – may have [2, 3]. As a way to mitigate the problem, during the World Conference on Disasters Risk Reduction in 2005, the United Nations launched the global campaign 'Hospitals Safe from Disasters'. This campaign was based on the commitment that hospitals must be safe in order to mitigate disasters damages and hospitals must continue functioning after a disaster at least three days after it occurred; therefore, the three criteria that a *safe hospital* must meet are: the protection of life, the protection of the investment and the protection of the function [4]. Thus, in order to certify a hospital as safe, a series of measurable structural (physical security) and non-structural (organizational) assessment criteria were developed. Such criteria are expressed in a *Safety*

DOI: 10.2495/SAFE-V8-N2-329-341

Index that do date has already been revised and published as a second version [5]. Furthermore, as governments began to focus on measuring the *safety* of hospitals, the academic sector in many countries began to investigate the *vulnerability* of hospitals to 'old risks' (i.e. vulnerability to floods and earthquakes), and more recently the risks and vulnerabilities associated with climate change. Complementing the research on hospital risks and vulnerabilities another field of research was developed; one that has to do with *risk management* in hospitals and has as main question how to overcome the barriers to manage risk in hospitals and the assessment of official risk reduction programs (see Table 1).

Table 1: Main subjects around the safety and vulnerability of hospitals as a field of research. *(Source: own elaboration based on the cited references.)*

Main questions	Main objectives	Countries	Authors (examples)
What are the impacts and consequences of disasters in hospitals?	To make emphasis on the need of a multidisciplinary approach to understand and solve the problem	Latin America and the Caribbean	ONU/CEPAL [3]
How safe is a hospital and what does safe mean?	Elaborating and applying a hospital safety index	Mainly Latin America and the Caribbean	WHO/PAHO [5]
How vulnerable is a hospital?	Seismic vulnerability assessment	India	Rautela *et al.* [6]
		Mexico	Moran-Rodríguez & Novelo-Casanova [7]
(*Risk and vulnerability assessment*)	Flood risk assessment	Sudan	Abbas & Routray [8]
	Assessment of disaster preparedness	Saudi Arabia	Bajow & Alkhalil [9]
		Japan	Mulyasari *et al.* [10]
	Climate change risks assessment	Australia and New Zealand	Loosemore *et al.* [11]
How well have risk reduction programs been implemented	Assessing implementation of activities aimed at reducing disaster risks	Cambodia, Lao PDR, the Philippines and Viet Nam	Geroy & Pesigan [12]
and how to overcome barriers to	Assessing implementation of training in emergencies	Mexico	Cruz-Vega *et al.* [13]
achieve implementation?	To identify barriers and strategies to increase safety	Iran	Yarmohammadian *et al.* [14]
(*Risk management*)	To explore resilience strategies	United Kingdom and Taiwan	Achour & Price [15]
	To identify and document successful experiences to makes hospitals safe	Argentina, Bolivia, Colombia, Chile, Ecuador, Paraguay and Perú	OPS/OMS [16]

It is in the last area of research that this paper inserts as a first approach to document the difficulties and success of risk management in Mexican hospitals, given the scarce literature on the subject in this country.

2 THE SAFE HOSPITALS PROGRAM IN MEXICO

The goal to achieve safe hospitals set out in the Hyogo Framework for Action in 2005– and ratified later in Sendai for 2015–2030 [17] – was accepted and signed by 168 countries, including Mexico. Thus, through the National System of Mexican Civil Protection, the Safe Hospital program in Mexico has been establishing since 2006 the instruments, procedures and mechanisms to evaluate and certify the level of safety of a hospital in areas subject to disaster risk in the entire Health System of Mexico. When evaluated, hospitals may be certified in the following three levels of safety [18]:

1. *Index Score A* means a good safety level but measures must continue to be taken in the medium and long term.
2. *Index Score B* is given to hospitals when short-term measures are required.
3. *Index Score C* is given to hospitals that have to undertake urgent measures immediately.

The purpose of assigning those scores is to know what kind of hospitals are available in case of disaster in order to apply the 'Triage' system (classification of patients immediately after the disaster): if the best equipped hospitals (essential hospitals) meet all safety requirements, they would be identified as 'red hospitals' because they will be *safe and able* to receive patients in a critical health condition and can perform complex operations that require special equipment and specialized medical staff. If hospitals are safe but have less infrastructure, they would be classified as 'green hospitals': they could receive patients who are discharged from the red hospitals but cannot perform complex medical tasks. Finally, if hospitals are safe but have just the basic medical equipment, they would be classified as "yellow hospitals" to receive patients for the most basic health functions in case of disaster.

To organize the evaluation of hospitals, the Mexican government divided the National Health System (made up of more than 3,000 hospitals) into three levels of complexity: low, medium and high complexity, the latter corresponding to hospitals that have the capacity to receive and operate victims of a disaster in a critical condition, because they have a certain number of operating rooms, beds, X-rays, Laboratory, ATLS-trained Doctors (Advance Trauma Life Support Course) and other requirements (see Fig. 1).

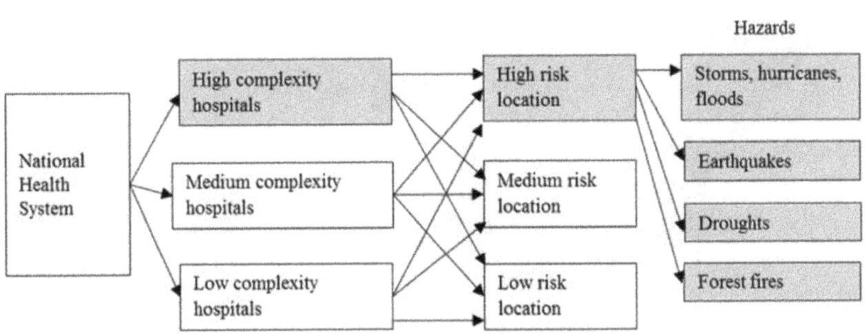

Figure 1: Organization of hospitals for evaluation into the Safety Hospitals Program in Mexico. Priority in grey. *(Source: own elaboration based on CVOED/IMSS.)*

Each one of these groups of hospitals (high, medium and low complexity) are located in territories with a different degree of risk: high risk, medium risk and low risk. The risk classification of hospitals was done by states, and it was based on a National Risk Atlas, recalling that in Mexico the exposure to hazards is differentiated and those risks still overlap: 41% of the national territory is exposed to hydrometeorological hazards (Storm, hurricanes and floods), 27% of the territory is exposed to earthquakes; 29% is exposed to drought and 37% is exposed to forest fires [19].

As far as we know, there is no an official or academic publication to show the results of the evaluation at the national level and there are just isolated data. It is known, for example, that from 2007 to 2013, 515 hospitals have been evaluated but the results of that classification (score of safety) is unknown [20]. In addition to these few data, in our bibliographical search there were very few publications about the vulnerability of hospitals in Mexico and no publication was found about the risk management in health facilities, so this work may show a first approach to the advances and obstacles of the implementation of the Safe Hospital Program in Mexico.

3 FLOOD RISK MANAGEMENT IN MEXICAN HOSPITALS

Given the regular presence of hurricanes in Mexico and the damages that result from them year after year, it was the intention of the authors of this paper to explore the effectiveness or the difficulty in achieving an acceptable level of safety in Mexican hospitals exposed to hydro-meteorological hazards in coastal zones. Such areas represent a very important tourist destination because of the natural and cultural heritage they have and the economic resources they generate, but they are also a vulnerable area for the population that lives there and whose livelihood depends directly from the touristic affluence.

For the selection of the case studies, we considered first the list of the hospitals classified as the ones of high complexity and high resolution according to the Safe Hospital Program Mexico´s. This list contained a total of 201 hospitals in 2011 [21]. From that list we selected then the hospitals of cities affected by hurricanes in the last years (see Fig. 2). Based on this information and the availability of human and financial resources to travel there, we selected the states of Colima, Jalisco, Chiapas and Tabasco. In each one of these hospitals interviews were carried out on site with emergency managers responsible for the Safe Hospitals Program.

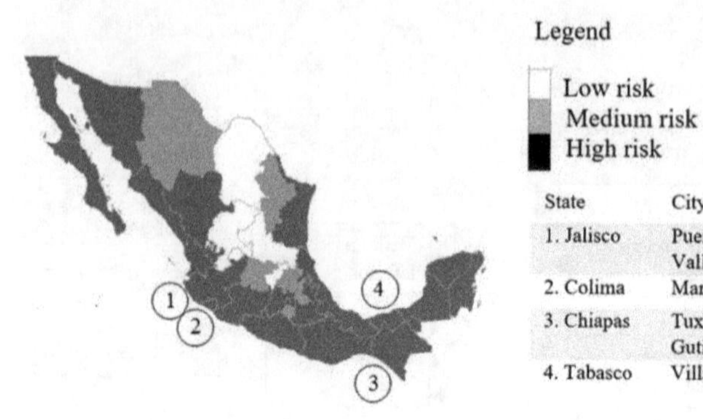

State	City	Hurricane
1. Jalisco	Puerto Vallarta	"Kenna", 2002
2. Colima	Manzanillo	"Jova", 2011
3. Chiapas	Tuxtla Gutiérrez	"Stan", 2005
4. Tabasco	Villahermosa	(Flood), 2007

Figure 2: Risk map of Mexico and some of the States affected by hurricanes or floods in the last years. *(Source: own elaboration based on CVOED/IMSS.)*

To carry out these interviews a questionnaire was formulated in order to document the existing barriers (internal or external) to the implementation of the program as well as the actions and strategies to overcome them; this method was based on the methodology proposed by Yarmohammadian *et al.* [14]. In the absence of a detailed list of hospitals evaluated and/or certified as safe hospitals, this question was the first one of the questionnaire. After obtaining the information in the field, the answers were classified by barriers and strategies *before*, *during* and *after* a flood.

A total of 10 hospitals were documented, representing four different States and four different institutions. We maintained the identity of the respondents anonymous so that no institution is compromised. The sample may seem small but the hospitals are the best equipped and there are just a few of them in each State. That is to say, in the majority of the States, just two or three hospitals are ranked as the ones of high complexity.

3.1 Main differences between hospitals

The first step to analyze the interviews was to make a differentiation between *hospitals with scarce works for certification* and *hospitals with advanced works for certification* (see Table 2). Among the first ones are those hospitals between 40 and 50 years old, for which it is no longer possible or would be too expensive a structural adaptation (stairs, elevators, ramps, etc.) in order to comply with the guidelines of the safe hospital Program; therefore for some of those hospitals the total abandonment of the building is already considered to be replaced by a new hospital even in another location. On the other group, hospitals that have been built more recently (from 5 to 20 years old) have an infrastructure that from the beginning allows them to adjust more easily to the security requirements; nevertheless this condition does not exempt them from having internal risks (i.e. derived from the equipment they handle), external risks, or even *new risks* such as organized crime. It was reported that in the case of some territories, it is common for the hospital to receive a patient that was victim of a criminal act, and who is in danger of being shot again at the hospital by a criminal group. This is a situation that puts at risk the medical staff. For this reason, one of the doctors is even submitting a normative protocol to ensure the life of the patient and the hospital staff.

A notable difference is the impact that both types of hospitals have received after a flood due to the presence of hurricanes or torrential rains. While better prepared hospitals have had only small leaks of water on ceilings or walls, unprepared hospitals had completely evacuated the building and taken patients to other hospitals. However, a common affectation to both types of hospitals during a flood is the lack of accessibility to the hospital because the streets surrounding it are flooded.

With regard to the management of the hospital, those responsible for the implementation of the Safe Hospital Program may be doctors, architects, or administrative personnel involved in civil protection. Their years of experience do not make a difference in both cases, although the knowledge about the Safe Hospital Program does. In certified hospitals or close to certification, some risk managers know the Program perfectly because they themselves are evaluators and they have evaluated other health facilities. This knowledge allows them to have a very clear vision of the tasks that must be done to increase the safety on a precise goals hierarchy and in a very timely manner (for example, budget needed to reinforce the hospital infrastructure, evidences of the drills performed, specialized equipment acquired, and equipment needed in the emergency area, dates of the next evaluation, etc.). On the other hand, in the less prepared hospitals, it was commented that they lack of evaluators or the staff is just next to receive the training to understand the functions and components of the same.

Table 2: Comparison of Mexican Hospitals according to requirements compliance in the Safe Hospitals Program. *(Source: Own elaboration based on interviews to hospital emergency managers.)*

Characteristics	Hospitals with scarce works for certification	Hospitals with advanced works for certification
Age of the building	Around 40–50 years old	Around 20 years
Internal risks reported	Explosion risk of hermetic equipment (boilers, autoclaves) and explosions risks derived from chemicals and energy supply accidents.	Fire risk, gas installations and chemicals (explosion risks).
External risks reported	Flood risk, earthquake risk, torrential rains, gas explosions, gas stations, epidemics, and the proximity to massive concentration of people in public spaces like parks and markets.	Proximity to a river, proximity to gas stations, proximity to boilers of supermarkets in the surroundings, geological risks, organized crime.
Type of damages as a consequence of past disasters	Flood into the hospital during hurricanes, Disruption of hospital activities or the complete evacuation of patients during floods. Reception of patients from other hospitals. Lack of accessibility from the outside.	Minimum damages in ceilings and walls (leaks). Lack of accessibility from the outside in case of flood.
Degree of knowledge of the Safe Hospital Program	Variable knowledge of the Program.	Very good knowledge of the Program.
Certified evaluators as part of the hospital staff	There are no certified evaluators in the hospital.	The hospital staff themselves are evaluators of the Safe Hospital Program.
Years of experience of the safety manager at the hospital	Variable	Variable
Field of expertise of the emergency manager	Variable	Variable
Detailed record of done and pendant tasks to compel with the Safe Hospital Program requirements	Diffuse goals to increase the safety of the hospital.	Clear record of the tasks done to increase the safety of the hospital with evidences shown to the interviewer.
Relocation plans	Relocation plan for the older hospitals.	No plans.

3.2 Barriers and Strategies for the Safe Hospital Program Certification

The age of the hospital, in addition to the – above mentioned – structural constraints also imposes barriers to the growth of the hospital for both permanent growth (i.e. new emergency rooms) and the emergency expansion that would allow the growth of the hospital, that is to say, a provisional growth to let the hospital expand temporarily to receive massive victims. The inability to expand the hospital also does not allow a growing number of patients to be treated, which keeps the hospital in permanent overpopulation, reducing the probability of receiving more patients in critical situations (see Table 3).

Administratively, the fact that the Safe Hospitals Program it is not completely mandatory, that impedes risk reduction actions from being a priority. In other cases there are already plans but they do not have a continuity due the change of administrations or they just require too much paperwork to be implemented. In addition staff must organize their time among many other responsibilities due to the lack of staff. As a consequence a reactive attitude towards risk reduction prevails over a preventive one.

Financial reasons were reported as another impediment to certifying the hospital. Since there is no specific budget for this program, this makes it difficult to the hospital to pay some technical works. An example of this is the technical opinions about the structural safety of the building. In addition, the lack of budget also makes it difficult to update some equipment or hire more staff or built more rooms to expand the hospital.

Finally, in almost all hospitals, a reactive and non-preventive attitude is one of the obstacles that prevent hospital staff from getting involved in the hospital's safety work. Certainly the economic part is fundamental to solve needs and increase the safety in hospitals, but one way to maximize the scarce economic resources is that investments in the emergency area (structural and non-structural) can also be planned to be useful for disaster preparedness, as stated by the respondents (see Table 4). An example of this is the expansion of the emergency area that considers decontamination requirements or the availability of medical equipment differentiated between children and adults in case of emergencies or disasters.

On the other hand, in spite of the economic restrictions that every institution faces, fortunately many nonstructural strategies were documented that are not necessarily expensive and can be done before, during or after a flood and just require a certain level of organization (see Tables 4 and 5). Examples of those measures are: risk mapping around the hospital that can be done by the same medical staff; emergency drills and drills for the use of the Triage system for the classification of patients in case of disaster; an internal plan for emergencies and above all the communication and coordination with civil protection authorities before and during a disaster. A fundamental aspect to implement the said strategies is the certification of evaluators of the Safe Hospital Program, which is a free course and that enables the medical staff to organize and execute this type of tasks.

For taking the evaluator course it is not necessarily a requirement to be a doctor or an emergency specialist, since each professional acquires an overview of the program but can deepen and contribute from his/her respective area of work: medicine, emergency, nursing, administration, engineering, construction or other fields. Indeed, the multidisciplinarity is a strength of the Program because each specialist focuses on his/her field of experience during an evaluation (i.e. the engineers focus on the structural aspects of the hospital).

An aspect that may help to solve some of the said barriers is the fact that the Safe Hospital Program was already established in the Mexican General Law on Civil Protection published in 2012 [22], and in the Official Mexican Norm NOM-016-SSA3-2012, that establishes the

Table 3: Barriers to comply with the Safe Hospitals Program in Mexican coastal zones. *(Source: Own elaboration based on interviews to hospital emergency managers.)*

Type of barriers	Barriers to increase safety
Technical and physical factors	At the time of construction of the hospital there were no zoning laws and the hospital was built too close to the river.
	Hospital was built before the paving of the streets. As a consequence, the street level is higher than the one of the hospital and water enters the building during heavy rains.
	Age of the building. Due to the period of construction the hospital does not have ramps to enable evacuation, it lacks of elevators and stairs of emergency, etc.
	Lack of space for regular activities and emergency expansion.
	Overpopulation of hospitalized patients (they function at 300% of the capacity). Lack of beds. In case of disaster, they cannot receive patients. Neighbour hospitals are in the same situation.
	Inaccessibility to the hospital in case of disasters.
	Reactive maintenance to equipment rather than preventive maintenance.
	Evacuation routes and identification of safe areas inside the hospital are not defined or not complete defined.
Institutional factors	There are other priorities and the Safe Hospital Program is not a mandatory program.
	In some cases lack of commitment from hospital authorities. Only reactive responses.
	The existing disaster preparedness remains just at a planning level.
	There are no continuity of plans because of the change of the authorities.
	The existing plans are focused on emergency and not on prevention.
	There are no enough staff hospital and more time is required to attend the program needs and requirements.
	Lack of experts to evaluate the hospital safety.
	Bureaucracy as an obstacle of the safety works.
	The requirements for certification are completed but there is still no an official request to certify.
	Lack of protocols for some kind of risks (i.e. how to deal with victims of organized crime.
Financial factors	There is no a specific budget to the Safe Hospital Program.
	Scarcity of financial resources.
	In a pre-evaluation, the cost of expert/technical opinions on structural, energy, electrical and gas installations is very expensive.
	There are no enough medical equipment.
	The existent equipment cannot be updated.
Human factors	There is no a ‘culture of risk’. Reactive instead of preventive responses.
	Lack of commitment from hospital staff.
	Interest in prevention actions just in the aftermath of the disaster.

Table 4: Strategies to reduce vulnerability to floods in Mexican hospitals in coastal zones and the main resources required. *(Source: Own elaboration based on interviews to hospital emergency managers.)*

Strategy to reduce vulnerability/Main resource required: F = Financial; O = Organizational	
Before the flood	
Structural strategies	
To have several entrances to the hospital to enable accessibility.	(F)
To expand the Emergency Room considering disaster preparedness too (for example a decontamination room with enough room for the patient stretcher).	(F)
A differential equipment in the emergency room for children and adult patients.	(O)
Adaptation of the street to serve as a dam to prevent water from entering the hospital.	(F)
Construction of a retaining wall close to rivers (with the support of the municipality).	(O)
Non-structural strategies	
Elaboration of hazards maps around the hospital (500 meters around) produced by the hospital staff.	(O)
To have and to follow a plan for –permanent- expansion/construction of the hospital.	(F)
To have a plan for emergency/temporal expansion with a triage system (an emergency expansion of at least 100%).	(O)
Commitment of the hospital staff according to their own experience and field of knowledge.	(O)
Permanent management to get resources and get financing from other institutions.	(O)
To motivate the staff to become evaluators.	(O)
Physicians with ATLS training (Advance Trauma Life Support), a special training for doctors.	(O)
Training courses for the hospital staff.	(O)
To make mandatory some of the activities to prepare the certification.	(O)
To define emergency brigades (fire brigade, for example).	(O)
Periodical and permanent drills of evacuation during flood and earthquakes (involving staff and patients).	(O)
Drill of patients-decontamination (chemical-biological).	(O)
Rope Descent/Rope rescue of patients (staff is at random selected for the drill once a year).	(O)
Cardiopulmonary resuscitation training and drills.	(O)
Plan of response during floods.	(O)
To perform Tsunami drill in coordination with authorities of civil protection and other hospitals of the region.	(O)
A precise diagnostics of the state of crucial machinery in case of disaster (water, energy, fire extinguishers, and so on).	(F)
To exchange risk management plans and strategies with other hospitals.	(O)
Before a heavy precipitation, portable water pumps are transferred to the hospital to suction water during flood.	(F)
Insurance of the hospital.	(F)
To elaborate an internal plan for emergencies and civil protection (for evacuation).	(O)

Table 5: Strategies to reduce vulnerability to floods in Mexican hospitals in coastal zones and the main resources required. *(Source: Own elaboration based on interviews to hospital emergency managers.)*

Strategy to reduce vulnerability/Main resource required: F = Financial; O = Organizational	
During the flood	
Non-Structural strategies	
Maintain communications with civil protection to evacuate staff and patients on time, if necessary.	(O)
If necessary, to evacuate patients and hospital staff to other hospitals.	(O)
Triage system after evacuation	(O)
The hospital staff organizes to take care of the evacuated patients.	(O)
Structural strategies	
If necessary, to rent another facility to give continuity of operations to the health care.	(F)
After the flood	
Structural strategies	
Technical inspection of the building structure after the flood to guarantee the safety.	(F)
To build a new hospital (as an extreme solution)	(F)

minimum characteristics of infrastructure and equipment for hospitals and specialized medical practices, and where the Checklist to apply the Safety Index appears [23].

4 CONCLUSIONS

The certification within the Safe Hospital Program represents a great challenge for any hospital in any institution because it involves a large number of aspects of a technical, administrative and financial nature; above all the implementation of the program demands a preventive attitude. The identified barriers should not be underestimated, as there are many other priorities in a highly complex hospital. In this sense, the institutional guidelines for making hospitals safe would require a more coercive character, so that recommendations to reduce vulnerability were actually implemented. At the same time, more financial resources are required to retrofit or, if necessary, replace inadequate hospitals with new infrastructure; it was observed that the age of hospitals is a strong limitation when new interventions or expansion is needed. Therefore, financial instruments for prevention and reinforcement should be a fundamental component in federal and institutional programs. The historical memory of the disaster also seems to be very important, as hospitals that have been impacted by disasters recently seem to be more aware of the need to prepare. Keeping in mind the occurrence of past damage could be an important point for the promotion of a culture of risk reduction. This reinforcement should involve the different levels of the hospital, both operational and managerial. Within the field of architecture there are many things that can be done too, both for the adaptation of old hospitals as well as for the creation of new paradigms of hospital design to accommodate the new requirements, not only those ones related to safety but also the possibilities to be flexible and adaptable to new and changing circumstances.

ACKNOWLEDGEMENTS

The authors of this paper are especially grateful to Dr. Sergio Germán Burgos Salcedo for having shared with us his great experience and deep knowledge on the safety of hospitals, particularly in the strategies part. We also thank Dulce Mara Rojas García, Adriana Jiménez Leyva and Francisco Javier Pérez Madrigal for their support in conducting some of the interviews. This article is a product of the research project IPN-SIP 20144135, funded by the Secretaría de Investigación y Posgrado del Instituto Politécnico Nacional.

REFERENCES

[1] Véliz-Pintos, R.A., Reflexiones de un pediatra ante la explosión de gas en un Hospital Materno Infantil de la Ciudad de México. *Revista Mexicana de Pediatría*, **82**(1), pp. 3–4, 2015. Available at: www.medigraphic.com/pdfs/pediat/sp-2015/sp151a.pdf. (accessed 16 December 2016).

[2] Organización Panamericana de la Salud, *Conferencia Internacional sobre Mitigación de Desastres en Instalaciones de Salud*. Available at: www.paho.org/disasters/index.php?option=com_content&view=article&id=664:international-conference-on-disaster-mitigation-in-heath-facilities&Itemid=924&lang=es (accessed 08 January 2015).

[3] ONU/CEPAL, *Impacto Económico de los Desastres Naturales en la Infraestructura de Salud*; ONU/CEPAL: México, 1996.

[4] International Strategy for Disaster Reduction/World Health Organization/World Bank, *Hospitals Safe from Disasters, Reduce Risk, Protect Health Facilities, Save Lives, 2008–2009 World Disaster Reduction Campaign,* ISRD/WHO/World Bank. Available at: www.unisdr.org/files/1347_wdrc20082009informationkit.pdf. (accessed 16 December, 2016).

[5] World Health Organization/Pan American Health Organization. *Hospital Safety Index: Guide for Evaluators* – 2nd ed., WHO: Geneva, 2015. Available at: www.who.int/hac/techguidance/hospital_safety_index_evaluators.pdf?ua=1. (accessed 07 September 2016).

[6] Rautela, P., Girish C.J. & Bhupendra, B., Seismic vulnerability of the health infrastructure in the Himalayan township of Mussoorie, Uttarakhand, India. *International Journal of Disaster Resilience in the Built Environment*, **2**(3), pp. 200–209, 2011. https://doi.org/10.1108/17595901111167088

[7] Morán-Rodríguez, S. & Novelo-Casanova, D.A., Seismic vulnerability assessment of Health facilities in Mexico City, Mexico. *WIT Transactions on The Built Environment*, vol 150, WIT Press, 2015. Available at: www.witpress.com/elibrary/wit-transactions-on-the-built-environment/150/33931. (accessed 05 September 2016).

[8] Abbas, H.B. & Routray, J.K., A semi-quantitative risk assessment model of primary health care service interruption during flood: Case study of Aroma locality, Kassala State of Sudan. *International Journal of Disaster Risk Reduction*, **6**, pp. 118–128, 2013. Available at: http://dx.doi.org/10.1016/j.ijdrr.2013.10.002. (accessed 12 November 2013).

[9] Bajow, N.A. & Alkhalil, S.M., Evaluation and analysis of hospital disaster preparedness in Jeddah. *Health*, **6**, pp. 2668–2687, 2014. Available at: http://dx.doi.org/10.4236/health.2014.619306. (accessed 11 September 2016).

[10] Mulyasari, F., Satomi, I., S., Prashar, S., Isayama, K., Basu, M., Srivastava, N. & Shaw, R., Disaster preparedness: looking through the lens of hospitals in Japan. *International Journal of Disaster Risk Science*, **4**(2), pp. 89–100, 2013. Available at: https://link.springer.com/article/10.1007/s13753-013-0010-1. (accessed 28 October 2014). https://doi.org/10.1007/s13753-013-0010-1

[11] Loosemore, M., Carthey, J., Chandra, V. & Chand, A.M., Climate change risks and opportunities in hospital adaptation. *International Journal of Disaster Resilience in the Built Environment*, **2**(3), pp. 210–221, 2011. Available at: http://dx.doi.org/10.1108/17595901111167097. (accessed 14 January 2015).

[12] Geroy, L.S.A. & Pesigan, A.M., Disaster risk reduction for health facilities in the Western Pacific Region. *International Journal of Disaster Resilience in the Built Environment*, **2**(3), pp. 268–277, 2011. https://doi.org/10.1108/17595901111167132

[13] Cruz-Vega, F., Loría-Castellanos, J., Hernández-Olivas, I.P., Franco-Bey, R., Ochoa-Ávila, C. & Sánchez-Badillo, V., Experiencia en capacitación en emergencias de la División de Proyectos Especiales en Salud, Instituto Mexicano del Seguro Social. *Cirugía y Cirujanos*, **84**(2), pp. 127–134, 2016. Available at: http://www.sciencedirect.com/science/article/pii/S0009741115002881. (accessed 13 September 2016). https://doi.org/10.1016/j.circir.2015.12.001

[14] Yarmohammadian, M.H., Atighechian, G., Shams, L. & Haghshenas, A., Are hospitals ready to response to disasters? Challenges, opportunities and strategies of Hospital Emergency Incident Command System (HEICS). *Journal of Research in Medical Sciences*, **16**(8), 1070–1077, 2011. Available at: www.ncbi.nlm.nih.gov/pmc/articles/PMC3263085/. (accessed 16 June 2014).

[15] Achour, N. & Price, A.D.F., Resilience strategies of healthcare facilities: present and future. *International Journal of Disaster Resilience in the Built Environment*, **1**(3), pp. 264–276, 2010.

[16] https://doi.org/10.1108/17595901011080869

[17] Organización Panamericana de la Salud/Organización Mundial de la Salud, *Comunidades Resilientes a través de Redes de Salud Seguras frente a Desastres en América del Sur*, OPS/OMS, 2012. Available at: www.cridlac.org/digitalizacion/pdf/spa/doc19246/doc19246-contenido.pdf. (accessed 13 September 2016).

[18] UNISDR (United Nations International Strategy for Disaster Reduction). *Sendai Framework for Disaster Risk Reduction 2015–2030,* 2015. Available at: www.preventionweb.net/files/43291_sendaiframeworkfordrren.pdf. (accessed 13 September 2016).

[19] Cruz, Felipe, Mexico: Applying the "Hospital Safety Index", 2009. Available at: http://safehospitals.info/index.php?option=com_content&task=view&id=119&Itemid=188. (accessed 28 January 2014).

[20] SEGOB/Banco Mundial, FONDEN: *Mexico's Natural Disaster Fund – A Review. The International Bank for Reconstruction and Development*/The World Bank. 2012, p. 3. Available at: www.proteccioncivil.gob.mx/work/models/ProteccionCivil/Almacen/fonden_resumen_ejecutivo.pdf. (accessed 12 June 2013).

[21] Centro Virtual de Operaciones en Emergencias y Desastres (CVOED). División de Proyectos Especiales en Salud. Dirección de Prestaciones Médicas, Instituto Mexicano del Seguro Social (IMSS). Available at: http://cvoed.imss.gob.mx/COED/home/normativos/DPM/capacitacion/curso_taller_evacuacion/3.%20HOSPITAL%20SEGURO.pdf. (accessed 18 January 2015).
[22] Coordinación de Protección Civil, Universo de hospitales clasificados como de alta complejidad y alto nivel resolutivo (rojos). Programa Hospital Seguro, México. Available at: www.proteccioncivil.gob.mx/work/models/ProteccionCivil/Resource/59/2/images/lhcr.pdf. (accessed 18 January 2015).
[23] Gobierno de México. Ley General de Protección Civil. Cámara de Diputados.
[24] Gobierno de México. Secretaría de Salud. Ley Norma Oficial Mexicana NOM-016-SSA3-2012, Que establece las características mínimas de infraestructura y equipamiento de hospitales y consultorios de atención médica especializada.

PRIORITIZATION OF HAZARDS BY MEANS OF A QFD- BASED PROCEDURE

MARA LOMBARDI & MARIO FARGNOLI
Faculty of Civil and Industrial Engineering, of the Sapienza University of Rome, Italy.

ABSTRACT

Despite the evolution of regulations in the field of occupational health and safety promoted in EU countries, the number of accidents and victims has not significantly decreased in recent years, especially in constructions and agriculture sectors, as underlined by official reports of the Italian Workers' Compensation Authority. Main reasons of such a situation are due to the characteristics of working activities in these sectors. The variety of operations, the frequent exchange of tasks among workers within the same company, the continuous change of workplaces, the frequent exchange of workers for the same activity (e.g. seasonal workers), and the workers' stress caused by seasonal jobs. For these reasons both risk assessment and safety management activities result in being more difficult than in other working sectors. Thus, it is important to provide methodologies and tools that allow companies to carry out these tasks more effectively. In such a context, the study proposed by Esra Bas in 2014 certainly represents an attempt to provide a supporting methodology for engineers engaged in risk assessment activities. This approach consists in the use of the Quality Function Deployment (QFD) method, and it is aimed at evaluating how specific tasks can be in relationship with specific hazards, which in turn are related to specific events, and finally at defining what preventive/protective measures can be introduced against those events. Based on this, we tried to further investigate such an approach, with the goal of providing an easier-to-use tool, which can be used in risk assessment activities of critical contexts as the agriculture one. With this aim in mind, a case study concerning the risk assessment of an agricultural machinery was carried out.

Keywords: agricultural equipment, house of quality, machinery safety, occupational safety, quality function deployment, risk assessment

1 INTRODUCTION

In last decades, regulations concerning occupational health and safety, as well as machinery safety have become more and more severe. Despite this, in most of EU countries the number of accidents and victims has not significantly decreased [1, 2]. Similarly, also in other developed countries such a phenomenon is perceived [3–6]. In particular, considering the Italian context, the occurrence of serious injuries and fatalities is very critical in the agriculture activities, where the compliance with new safety requirements results in being more difficult, due to the specific characteristics of the sector [7]. Actually, it has to be underlined that difficulties related with the implementation of safety measures are numerous, in particular for small and medium sized companies (SMEs), which can hardly afford additional costs for the compliance with safety requirements. SMEs represent more than the 80% of enterprises operating in the agricultural sector in Italy, and most of them are family-owned companies. Moreover, the number of part-time workers is also consistent.

Most of serious accidents in this sector occur while using self-propelled machines, and in particular tractors [8–11], because of:

- The obsolescence of a large number of machinery still used nowadays, which need to be adapted to the latest safety requirements;
- The lack of knowledge of safety rules when using these machines, including correct procedures when carrying out maintenance activities and setting-up operations.

DOI: 10.2495/SAFE-V8-N2-342-353

Thus, it is clear that both risk assessment and safety management activities result in being more difficult than in other working sectors. For these reasons, the development of methodologies and tools that allow companies to carry out these tasks more efficaciously is discussed in literature, e.g. in Refs. [12–16]. Also the involvement of both users and equipment providers of agriculture machinery is important to properly analyse risk factors and the effectiveness on preventive and protective measures adopted, since it is recognized that most of occupational accidents are caused by unsafe conditions and unsafe behaviours [14, 17–21]. Therefore, performing risk assessment in a sequential manner (i.e. cause-effect analysis) is insufficient to take into account such a complexity. As argued by Ref. [22], a holistic approach is needed to include interactions among different agents (operators, technical systems, working environment). In such a context, the study proposed by Ref. [1] certainly represents an attempt to provide a supporting methodology for engineers engaged in risk evaluations. On these considerations, to further investigate the use of the Quality Function Deployment method in safety risk management, the paper presents a practical case study concerning the risk analysis of an agricultural tractor taking into account its multiple uses options.

The remainder of the paper is organized as follows. In Section 2 a background analysis concerning the occupational accidents trends and safety issues in the agricultural sector is reported. Then Section 3 describes main characteristics of the proposed research approach, while in Section 4 its implementation to a practical case study is shown. Hence results obtained are discussed in Section 5, while Section 6 addresses conclusions and further work.

2 BACKGROUND ANALYSIS

In the last years standards and regulations concerning occupational safety have become increasingly strict, following the action of the European Union, aimed at making the framework of safety requisites to protect both consumers and professionals from unsafe products more adequately. In such a context, the New Legislative Framework (NLF) is a clear example of this trend, which is oriented at guaranteeing an equal emphasis on product related requirements to be met when products are placed on the market, as well as on enforcement safety aspects during the whole life-cycle of products [23].

Nevertheless, as mentioned above, the number of accidents has not significantly decreased, especially considering small and medium sized enterprises operating in agriculture and forestry sectors. Latest data concerning the occurrence of occupational accidents in the sector of agriculture [24] show that the problem is very relevant in Italy, considering both its absolute value, as well as the number of accidents occurred in agricultural activities in comparison with the ones registered in other sectors (Table 1). As a matter of facts, these data need to be read considering the number of workers operating in the various sectors, i.e. the frequency factor (number of accidents per 1000 of employees). In this case, in 2015–2016 the frequency factor in industry and services was ca. 4,36, while in agriculture it is more than twice this number.

Table 1: Official statistics on occupational accidents per sector *(Source: INAIL, 2017)*.

Sector	**2011**	**2012**	**2013**	**2014**	**2015**	**2016**
Industry and service	79%	78%	77%	77%	78%	78%
Agriculture	6%	6%	6%	6%	6%	6%
Public	15%	16%	17%	17%	17%	16%
Total	**817.778**	**745.572**	**694.969**	**663.493**	**632.665**	**636.812**

A more detailed analysis of these data shows that in agriculture also the number of permanent injuries is quite large and the most dangerous activities (considering the number of permanent injuries and fatalities which occurred) are the ones which involve the use of machines and mechanical equipment, and in particular the use of tractors (Table 2). Furthermore, also accidents occurred on the road were considered: in the period January 2016–July 2016, according to ASAPS [25], 243 cases involving tractors were registered, causing 106 fatalities among tractors' drivers and passengers, as well as 162 people injured.

Taking into account data provided by INAIL, the Italian Workers' Compensation Authority [26], it was possible to further analyse the types of accidents occurred while using a tractor during in field working activities. As shown in Table 3, most of them are due to roll-over, running-over and falling.

The analysis of the way events occurred showed that this type of accidents are related to different activities, such as tillage, harvesting and pesticide application, or when the tractor is used as a power unit (i.e. due to an unsafe us of the tractor's power take-off (PTO)). The main reasons of such a situation can be found in the following aspects:

- A large number of tractors used in agricultural activities are not up-dated following the recent development of safety standards and regulations (the average age is more than 20 years old in accordance with data provided by UNACMA (National Union of Agricultural Machinery Dealers)).
- Almost 50% of operators are part-time workers (i.e. seasonal workers, hobbyists), whose expertise and knowledge of safety procedures is poor (i.e. lack of proper use of PPEs (Personal Protection Equipment), compliance with latest laws and regulations, proper maintenance activities).

Table 2: Trend of fatal accidents in agriculture *(Source: INAIL, 2017).*

Year	**Fatal accidents in agriculture**	**Fatal accident involving tractors**	**Percentage of fatal accident involving tractors**
2013	199	121	60.8 %
2014	189	121	64.0 %
2015	205	137	66.8 %
2016	197	114	69.5 %

Table 3: Types of fatal accidents involving tractors *(Source: INAIL, 2017).*

Type	**2010**	**2011**	**2012**	**2013**	**2014**	**2015**	**2016**
Roll-over	116	80	85	89	100	106	90
Running-over	9	13	7	13	9	15	9
Falling	8	11	6	7	10	10	8
Collision		5	7	3		4	1
Others	2	18	8	9	2	2	6
Total	**135**	**127**	**111**	**121**	**121**	**137**	**114**

If we consider also difficulties related to the great variety of activities, as well as the continuous change of workplaces, it appears clear why risk assessment and safety management activities in agriculture result in being more difficult than in other working sectors. For these reasons, it is important to provide methodologies and tools that allow companies to carry out these tasks more effectively.

3 MATERIALS AND METHODS

The proposed approach is based on the use of the Quality Function Deployment (QFD) method [27], following research hints provided by Bas [1].

QFD is widely recognized as one of the most effective tools for the design and development of products and services. The core of the method relies on the House of Quality (HoQ), i.e. a set of matrices that allows engineers to translate the so-called 'Voice of Customers' (VoCs) into engineering parameters [28]. In Fig. 1 a scheme of the HoQ is shown, where customer requirements ('Whats') are matched with engineering characteristics ('Hows') by means of a relationship matrix. Other parts of the HoQ are two matrices, used to assess relationships among themselves of the Whats and the Hows respectively. The output of the HoQ consists in the assessment of engineering characteristics and in their quantitative definition (i.e. the so-called 'How-muches').

In the last years, the use of QFD has been extended to the improvement of different properties related to products and services, such as: environmental sustainability, reliability, maintainability, etc., while its use in the field of health and safety, and risk assessment has been scarcely investigated as pointed out by some authors [29]. This is due to the fact that main issues and targets concerning safety are defined by laws and regulations (e.g. directive 2006/42/EC in the case of machinery). Hence, whenever a higher level of safety can be achieved it becomes the standard level. Based on this, most of studies on the application of QFD in health and safety field concern the assessment of safety requirements as target values

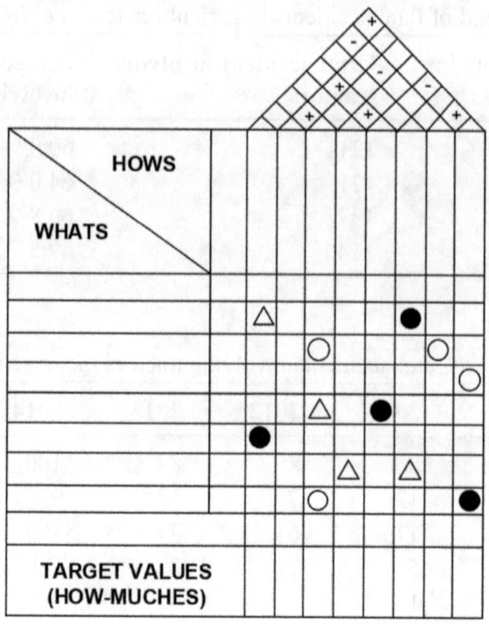

Figure 1: Scheme of the House of Quality (HoQ).

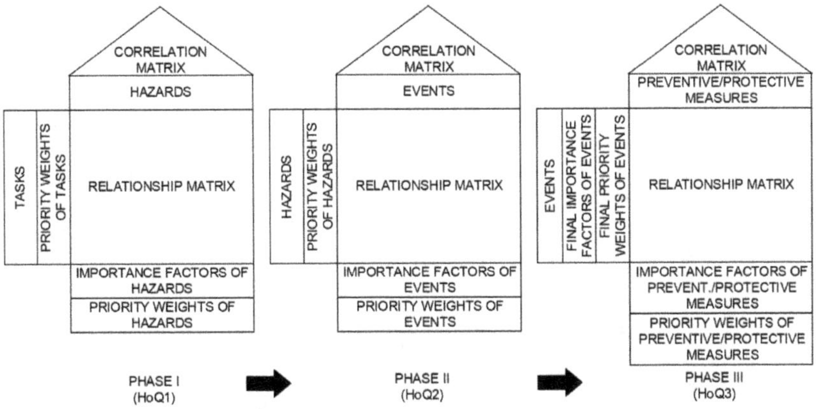

Figure 2: Scheme of the QFD for Safety adapted from Ref. [1].

[7, 29, 30]. A different approach is the one proposed by both Abubakar [8] and Bas [1], who used the 'mechanism' of the HoQ as a tool to assess cause-effects relationships. In particular, the latter presented a three-phase procedure (Fig. 2):

- Phase I – evaluation of the relationships between tasks and hazards;
- Phase II – evaluation of the relationships between hazards and events;
- Phase III – comparison of events with preventive and protective measures.

Such an approach (Fig. 2) was discussed through its hypothetical application in two different contexts: a project in construction industry and an IT firm belonging to SMEs category.

Based on this, we tried to apply it in practice by means of a real case study in the agriculture field with the aim of verifying its effectiveness as a risk assessment tool.

4 CASE STUDY

More in details, the risk assessment concerned agricultural activities carried out by a small company that produces wheat using a wheeled cabin tractor and several agricultural machinery and equipment (e.g. trailers, ploughs, harrows, etc.). Main technical characteristics of the tractor are the followings:

- Power: 80 kW.
- Weight: 4000 kg (un-ballasted).
- Axels: 4WD front axle.
- Wheelbase: 2422 mm.
- Track width (min/max): 1625–2025 mm.

Firstly, the set of tasks of a given project and the possible hazards that may arise from performing these tasks are analyzed. The definition of tasks carried out in collaboration with the company's experts, while hazard identification and assessment was performed with the support of both experts belonging to INAIL, the Italian Workers' Compensation Authority, as well as experts belonging to UNACMA (the Italian Association of Dealers of Agricultural Machinery). The set of the T_m tasks and the list of the h_n hazards were defined, and their relationships were quantified by means of the HoQ_1 (Table 4).

Table 4: Correlation matrix of the HoQ1.

HoQ₁	Priority weights of tasks	h1 – Machinery mobility	h2 Moving elements	h3 – thermal hazards	h4 – stability	h5 – Environment (dust, heat, gradient)	h6 – Noise	h7 – Vibrations	h8 – Ergonomic (Access, Posture)	h9 – Ergonomic (Commands)	h10 – Combination (stress)
T1 – Setting	0.1	0	3	9	0	0	0	0	0	9	9
T2 – Trailing	0.1	9	0	0	9	0	3	9	9	0	9
T3 – Ripping	0.1	3	3	0	3	3	3	9	3	3	0
T4 – Plowing	0.1	3	3	0	3	3	3	9	3	3	0
T5 – Tilling	0.1	3	3	0	3	3	3	9	3	3	0
T6 – Shredding	0.1	3	3	0	3	3	3	9	3	3	0
T7 – Sowing	0.1	3	3	0	3	3	3	3	3	3	0
T8 – Mowing	0.1	3	9	0	0	9	3	3	3	9	0
T9 – Raking and tedding	0.1	9	9	0	0	9	0	3	3	3	3
T10 – baling	0.1	9	3	0	0	3	0	9	3	9	3
Importance factors of hazards		4.50	3.90	0.90	2.40	3.60	2.10	6.30	3.30	4.50	2.40
Priority weights of hazards		0.652	0.565	0.130	0.348	0.522	0.304	0.913	0.478	0.652	0.348

Similarly, the events that the hazards may cause and their mutual relationship were quantified using the HoQ2 (Table 5).

In the HoQ3, the final importance factors and priority weights of events, as well as the hierarchy of preventive/protective measures were defined (Table 6) and assessed (Table 7). It has to be pointed out that the relationships between the events and preventive/protective measures were quantified by means of the following formula, which provides the final priority weights of events (w_i):

$$w_i = (o_{i,nf} \cdot s_{i,nf} + o_{i,f} \cdot s_{i,f}) \qquad (1)$$

where:

$o_{i,nf}$ represents the probability of occurrence of event e_i that leads to a non-fatal injury;
$s_{i,nf}$ represents the expected severity of event e_i that leads to a non-fatal injury;
$o_{i,f}$ represents the probability of occurrence of event e_i that leads to a fatal injury; and
$s_{i,f}$ represents the expected severity of event e_i that leads to a fatal injury.

Table 5: Correlation matrix of the HoQ2.

HoQ$_2$	Priority weights of hazards	E1 – Contact with cutting parts	E2 – Contact with rotating parts	E3 – Contact with high temperature parts	E4 – Loss of stability	E5 – Contact with ejected objects	E7 – Exposure to vibrations	E6 – Exposure to noise	E8 – Slipping, falling	E9 – Unintended movements	E10 – Loss of control	E11 – Impacts with falling objects
H1 – Machinery mobility	0.652	0	3	0	9	0	3	3	9	3	3	0
H2 – Moving elements	0.565	3	9	0	0	0	3	3	0	0	0	0
H3 – Thermal hazards	0.130	0	0	9	0	3	0	0	0	0	0	0
H4 – Stability (machinery)	0.348	0	0	0	9	3	0	0	3	3	9	0
H5 – Environment (Dust, etc.)	0.522	0	0	0	3	3	0	0	3	3	3	3
H6 – Noise	0.304	0	0	0	0	3	0	9	0	0	3	0
H7 – Vibrations	0.913	0	0	0	0	3	9	0	0	0	3	0
H8 – Ergonomic (Posture, etc.)	0.478	0	0	0	0	9	0	0	9	0	0	0
H9 – Ergonomic (Commands)	0.652	3	3	0	0	9	3	3	3	9	9	0
H10 – Combination (stress)	0.348	9	9	0	0	3	0	0	3	3	3	0
Preliminary importance factors of events		6.78	12.13	1.17	10.57	17.87	13.83	8.35	15.78	11.48	17.22	1.57
Preliminary priority weights of hazards		0.813	1.453	0.141	1.266	2.141	1.656	1.000	1.891	1.375	2.063	0.188

Table 6: Set of events.

Set of events	
P1 – PTO protection	P11 – Seat's restraint system
P2 – Transmission belt protection	P12 – Mechanical block/unblock of parking
P3 – Protection of the gearbox	P13 – On/Off locking differential system
P4 – Access handles	P14 – Safety instructions
P5 – ROPS	P15 – PPEs
P6 – Anti-vibration seat	P16 – Shredders' guards
P7 – Adequate muffler	P17 – Mower's guard
P8 – Cabin filters	P18 – FOPS
P9 – Seat's 'dead-man switch'	P19 – Acoustic warning devices
P10 – Seat's adjustment leverages	

Table 7: Excerpt of the correlation matrix of the HoQ3.

HoQ₃	Final priority weights of events	P1 – PTO protection	P2 – Transmission belt protection	P3 – Protection of the gearbox	P4 – Access handles
E1 – Contact with cutting parts	0.9995	0	0	0	0
E2 – Contact with rotating parts	1.7875	9	9	0	0
E3 – Contact with high temperature parts	0.1670	0	0	9	0
E4 – Loss of stability	2.0400	0	0	0	0
E5 – Contact with ejected objects	0.0121	0	0	0	0
E7 – Exposure to vibrations	0.0000	0	0	0	0
E6 – Exposure to noise	0.0000	0	0	0	0
E8 – Slipping, falling	0.1657	0	0	0	9
E9 – Unintended movements	0.0194	0	0	0	0
E10 – Loss of control	0.1546	0	0	0	0
E11 – Impacts with falling objects	0.0005	0	0	0	0
Importance factors of preventive/protective measures		16.09	16.09	1.50	1.49
Priority weights of preventive/protective measures		0.541	0.541	0.051	0.050

Differently from Ref. [1], in this case the calculation of the final priority weights of events did not consider costs related to events were not considered since this was not a goal of the study. Moreover, as for the calculation of the final importance factors and priority weights of events, the probability of occurrence of both fatal and non-fatal injuries, as well as the expected severity of injuries were estimated taking into account data provided by INAIL [26].

5 DISCUSSION OF RESULTS

As it can be seen in Fig. 3, the most relevant protective/preventive measures concern precisely:

1. P17 – Mower's guards;
2. P11 – Seat's restraint system;
3. P16 – Shredders' guard;
4. P14 – Safety instructions (tractor and equipment)
5. P1 – PTO protection and P2 – Transmission belt protection.

The priority of interventions reflects the *status quo* of the tractor and other equipment used by the company, e.g. in Fig. 4 the picture of the tractor's seat is shown, where the absence the seat belt can be noted. Thus, the present study confirmed the effectiveness of the use of the approach proposed by Bas [1] to perform risk assessment activities. The use of a focus group of experts (including users, machinery manufacturers/retailers and authorities) allowed us to

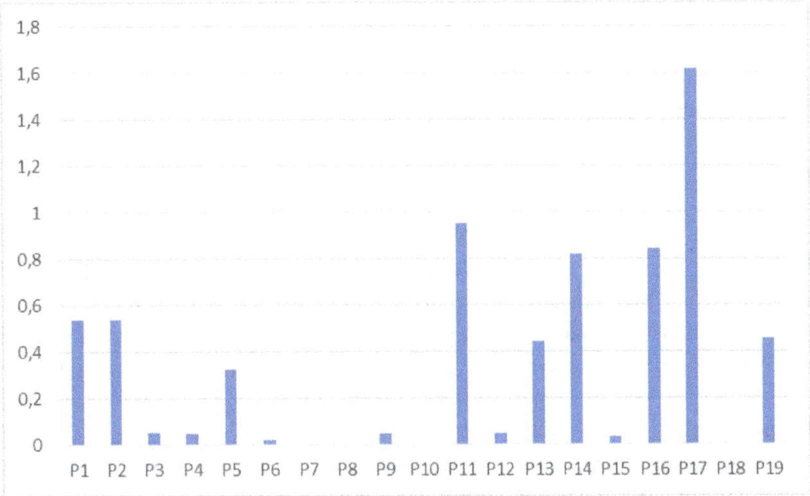

Figure 3: Priority weights of preventive/protective measures.

Figure 4: Details of the operator seat in the tractor's cabin.

identify and estimate all the relationships between tasks and hazards, hazards and events, as well as events and preventive/protective measures. In other words, the HoQ mechanism enables engineers to carry out a cause-effect analysis that provides an assessment of each interrelationship between tasks and hazards, hazards and events, and events compared with preventive/protective possible interventions.

Nevertheless, some limitations should be considered as well. Firstly, the assessment of the effects of the exposure to noise and vibrations could not be considered due to the lack of specific information. Secondly, the proposed methodology is based on the use of official statistics: the more these data are accurate the more the methodology can provide precise results. Then, it has to be noted that the relationship with cogent safety requirements is not provided, thus an additional effort is needed to provide solutions that are in compliance with current safety laws.

Finally, from a more general point of view, an additional limitation can be found in the risk of an incorrect assessment of qualitative characteristics and attributes, which is related to the use of the HoQ (as pointed out, for example, by Refs. [31–34]). To solve such a problem, as suggested by Ref. [1], the use of Fuzzy Logic could be considered.

6 CONCLUSIONS AND FURTHER WORK

In this study, a specific Safety-QFD methodology was implemented through its application to agricultural activities. More in details, the House of Quality was used following the approach proposed by Ref. [1] as a risk assessment tool, with the aim of identifying possible risks related with the use of a tractor and several equipment to carry out usual works in wheat farming.

Results achieved show that such a methodology can be used for the risk assessment effectively, allowing engineers to obtain the priority of interventions aimed at increasing the safety level of the working activities. Although these positive outputs, further research work is needed to exceed the above-mentioned limitations. In particular, a comparison with the traditional risk assessment approach suggested by the ISO 12100:2010 standard is foreseen. In addition, an easier-to-use tool for the assessment of safety costs, as the one proposed by Ref. [35] is currently being analysed.

ACKNOWLEDGEMENTS

Authors wish to thank Vincenzo Laurendi, PhD, and Daniele Puri, INAIL's researchers, for their support in the case study development.

REFERENCES

[1] Bas, E., An integrated quality function deployment and capital budgeting methodology for occupational safety and health as a systems thinking approach: the case of the construction industry. *Accident Analysis & Prevention*, **68**, pp. 42–56, 2014. https://doi.org/10.1016/j.aap.2013.10.005

[2] Eurostat, 2016. Available at: http://ec.europa.eu/eurostat/statistics-explained/index.php/Accidents_at_work_statistics. (Accessed 13 April 2017).

[3] Frank, A. L., McKnight, R., Kirkhorn, S. R. & Gunderson, P., Issues of agricultural safety and health. *Annual Review of Public Health*, **25**, pp. 225–245, 2004. https://doi.org/10.1146/annurev.publhealth.25.101802.123007

[4] Caffaro, F., Mirisola, A. & Cavallo, E., Safety signs on agricultural machinery: pictorials do not always successfully convey their messages to target users. *Applied Ergonomics*, **58**, pp. 156–166, 2017. https://doi.org/10.1016/j.apergo.2016.06.003

[5] Douphrate, D.I., Stallones, L., Lunner Kolstrup, C., Nonnenmann, M.W., Pinzke, S., Hagevoort, G.R. & Jarvie, P., Work-related injuries and fatalities on dairy farm operations—a global perspective. *Journal of Agromedicine*, **18**(3), pp. 256–264, 2013. https://doi.org/10.1080/1059924x.2013.796904

[6] Burgus, S. & Duysen, E., Identifying topics and dissemination methods for agricultural safety and health messages. *Safety*, **3**(3), p. 3, 2017. https://doi.org/10.3390/safety3010003

[7] Fargnoli, M., Vita, L., Gattamelata, D., Laurendi, V. & Tronci, M., A reverse engineering approach to enhance machinery design for safety. *Proceedings of the 12th International Design Conference - DESIGN 2012*, pp. 627–636, 2012.
[8] Abubakar, M.S.A., Ahmad, D. & Akande, F.B., A review of farm tractor overturning accidents and safety. *Pertanika Journal of Science & Technology*, **18**(2), pp. 377–385, 2010.
[9] Mayrhofer, H., Quendler, E. & Boxberger, J., Occupational incidents with self-propelled machinery in Austrian agriculture. *Journal of Agromedicine*, **18**(4), pp. 359–367, 2013.
https://doi.org/10.1080/1059924x.2013.827997
[10] Cavallo, E., Ferrari, E., Bollani, L. & Coccia, M., Attitudes and behaviour of adopters of technological innovations in agricultural tractors: A case study in Italian agricultural system. *Agricultural Systems*, **130**, pp. 44–54, 2014.
https://doi.org/10.1016/j.agsy.2014.05.012
[11] Li, Z., Mitsuoka, M., Inoue, E., Okayasu, T. & Hirai, Y., Development of stability indicators for dynamic Phase I overturn of conventional farm tractors with front axle pivot. *Biosystems Engineering*, **134,** pp. 55–67, 2015.
https://doi.org/10.1016/j.biosystemseng.2015.03.016
[12] Myers, M.L., Cole, H.P. & Westneat, C., Injury severity related to overturn characteristics of tractors. *Journal of Safety Research*, **40**, pp. 165–170, 2009.
https://doi.org/10.1016/j.jsr.2009.02.007
[13] Myers, J.R. & Hendricks, K.J., Agricultural tractor overturn deaths: assessment of trends and risk factors. *American Journal of Industrial Medicine*, **53**(7), pp. 662–672, 2010.
https://doi.org/10.1002/ajim.20775
[14] Fargnoli, M., Laurendi, V. & Tronci, M., Design for safety in agricultural machinery. *Proceedings of the 11th International Design Conference - DESIGN 2010*, pp. 701–710.
[15] Donham, K.J. & Thelin, A., *Prevention of Illness and Injury in Agriculture. Agricultural Medicine: Rural Occupational and Environmental Health*, Safety, and Prevention, John Wiley & Sons, pp. 503–550, 2016.
[16] Lombardi, M., Guarascio, M. & Rossi, G., The management of uncertainty: model for evaluation of human error probability in railway system. *American Journal of Applied Sciences*, **11**(3), pp. 381–390, 2013.
https://doi.org/10.3844/ajassp.2014.381.390
[17] Rautiainen, R.H., Ledolter, J., Donham, K.J., Ohsfeldt, R.L. & Zwerling, C., Risk factors for serious injury in Finnish agriculture. *American Journal of Industrial Medicine*, **52**(5), pp. 419–428, 2009.
https://doi.org/10.1002/ajim.20688
[18] Taattola, K., Rautiainen, R.H., Karttunen, J.P., Suutarinen, J., Viluksela, M.K., Louhelainen, K. & Mäittälä, J., Risk factors for occupational injuries among full-time farmers in Finland. *Journal of Agricultural Safety and Health*, **18**(2), pp. 83–93, 2012.
https://doi.org/10.13031/2013.41348
[19] Liu, H.T. & Tsai, Y.L., A fuzzy risk assessment approach for occupational hazards in the construction industry. *Safety Science,* **50**, pp. 1067–1078, 2012.
https://doi.org/10.1016/j.ssci.2011.11.021
[20] Kogler, R., Near accidents with agricultural vehicles, machinery and equipment in Austria in the year 2013. *Agricultural Engineering International: CIGR Journal*, **17**(1), pp. 141–157, 2015.

[21] Kogler, R., Quendler, E. & Boxberger, J., Occupational accidents with agricultural machinery in Austria. *Journal of Agromedicine*, **21**(1), 61–70, 2016. https://doi.org/10.1080/1059924x.2015.1075451
[22] Underwood, P. & Waterson, P., Systemic accident analysis: examining the gap between research and practice. *Accident Analysis & Prevention*, **55**, pp. 154–164, 2013. https://doi.org/10.1016/j.aap.2013.02.041
[23] European Union. The 'Blue Guide' on the implementation of EU products rules 2016. *Official Journal of the European Union, C 272/1*, **59**, 2016.
[24] INAIL, 2016. 'Informo' database concerning occupational accidents causing fatal and serious injuries. Available at: https://appsricercascientifica.inail.it/getinf/informo/home_informo.asp. (Accessed 13 April 2017).
[25] ASAPS, Database of the Association for traffic safety, 2016. Available at: http://www.asaps.it/. (Accessed 13 April 2017).
[26] INAIL, Osservatorio INAIL sugli infortuni nel settore agricolo e forestale (National observatory on accidents in agricultural and forestry activities), 2017.
[27] Akao, Y., *Quality Function Deployment: Integrating Customer Requirements into Product Design*. Productivity Press, Cambridge, MA, 1990.
[28] ReVelle, J.B., Moran, J.W. & Cox, C.A., *The QFD Handbook*, John Wiley & Sons, Inc., 1998.
[29] Fargnoli M. & Sakao T., Uncovering differences and similarities among Quality Function Deployment based methods in Design for X - benchmarking in different domains. *Quality Engineering*, pp. 1–23, 2016. https://doi.org/10.1080/08982112.2016.1253849
[30] Moores, B.M., Radiation safety management in health care – The application of Quality Function Deployment. *Radiography*, **12**(4), pp. 291–304, 2006. https://doi.org/10.1016/j.radi.2005.07.011
[31] Wang, X., Wang, L., Xu, X. & Ji, P., Identifying employee turnover risks using modified quality function deployment. *Systems Research and Behavioral Science*, **31**(3), pp. 398–404, 2014. https://doi.org/10.1002/sres.2282
[32] Carnevalli, J.A. & Miguel, P.C., Review, analysis and classification of the literature on QFD-Types of research, difficulties and benefits. *International Journal of Production Economics*, **114**(2), pp. 737–754, 2008. https://doi.org/10.1016/j.ijpe.2008.03.006
[33] Wang, C.H. & Chen, J.N., Using quality function deployment for collaborative product design and optimal selection of module mix. *Computers & Industrial Engineering*, **63**, pp. 1030–1037, 2012. https://doi.org/10.1016/j.cie.2012.06.014
[34] Vinayak, K. & Kodali, R., Benchmarking the quality function deployment models. *Benchmarking: An International Journal*, **20**(6), pp. 825–854, 2013. https://doi.org/10.1108/bij-07-2011-0052
[35] Fargnoli, M. & Pighini, U., SCRM: a new Design Tool for improving Safety Level of Mechanical Systems. *Proceedings of the 14th International Conference on Engineering Design, ICED 03*, pp. 553–554 (exec.summm.), 2003.

DETERMINING SIZE FRACTIONS OF AN AEROSOL IN VIEW OF NEW DEFINITIONS IN POLISH REGULATION

PRZEMYSLAW OBERBEK & BEATA KACZOROWSKA
Central Institute for Labour Protection – National Research Institute, Warsaw, Poland.

ABSTRACT

A Polish regulation of the Minister of Labour and Social Policy on the value of the maximum concentrations of harmful chemical and dust factors in the workplace introduced new definitions of aerosols, which is "inhalable fraction and respirable fraction in place of the total dust and respirable dust". Due to the modification of the definitions of aerosol fractions, it is necessary to verify methods for determining the inhalable and respirable fraction of the aerosol. The purpose of the studies leading to the determination of the inhalable fraction and the respirable fraction of the aerosol is to obtain validated and reliable exposure information to compare the exposure to the limit value. The most reliable results of occupational exposure assessment are obtained using an individual dosimetry method. The research was carried out at seven selected workstations where there was emitted, for example, wood dust or coal dust. Dust samples were collected in a breathing zone on a specially constructed stand containing a set of sampling heads along with individual aspirators. At each of the investigated workstations, samples of the inhalable fraction and respirable fraction of the dust were collected simultaneously using different sets of sampling heads capable of sampling with varying volume air flow. These studies were conducted to compare the various samplers currently available on the Polish market. IOM was used as a reference personal sampler. By analyzing the results, it was found that the IOM sampler at each workstation obtained similar values (at least four IOM heads were used in one test). The biggest discrepancies can be seen with the GSP 3.5 sampler.

Keywords: aerosols, chemical hazards, coal dust, dust hazards, individual dosimetry method, inhalable fraction, occupational exposure assessment, respirable fraction, wood dust

1 INTRODUCTION

According to the Central Statistical Office (pol. *Główny Urząd Statystyczny*, GUS) data on persons employed in hazardous conditions, among the work-related factors, in 2015 there were 59.8 thousand persons exposed to industrial dust [1] (Zgierska *et al.*, 2016). The term 'industrial dust' is defined as any aerosol the dispersed phase of which is composed of solid particles (grains) within the diameter range of 0.1 to 100 μm, generated by activities employed in manufacturing processes. For the purpose of dust particle size determination, the following terms are employed:

- equivalent grain diameter, the diameter of a spherical particle with density identical to the density of investigated dust particle, which exhibits equal falling speed in non-turbulent air as the investigated dust particle,
- aerodynamic diameter - the diameter of a sphere with the density of 1 g/cm^3 which exhibits equal falling speed in non-turbulent air as the investigated dust particle,
- projection diameter - the diameter of a circle, the area of which is equivalent to the area of the investigated particle projection on the observation plane.

Particulate pollutant emissions are a consequence of most industrial processes. Among processes contributing the most to particulate pollutant emissions are: grinding, milling, sieving, transport, mixing and extraction of resources. The most harmful high-dispersion dusts are created in large concentrations during the processes of sharpening, grinding and polishing.

DOI: 10.2495/SAFE-V8-N2-354-366

Health effects of dust exposure usually become apparent after a latent period in the form of a variety of respiratory diseases, including pathological connective tissue growth (contributing to pneumoconiosis) and cancer [2] (Więcek E., 2011).

In the Ordinance of the Minister of Labour and Social Policy of 6 June 2014 on highest permissible levels of concentration and intensity of agents harmful to health in the working environment Polish (Journal of Laws 2014, item 817), considering the recommendations of the PN-EN 481:1998 [3] standard, modified definitions of aerosol fractions were introduced - notably the inhalable and the respirable fractions. Such actions are in accordance with the European approach to the determination of aerosol fractions in the working environment, shown in the EN 481:1993 [4] standard, where air sampling principles were introduced, basing on the premise that particles which can pass to the human body are in the breathing zone (around the worker's nose and mouth). In the EN:481:1993 standard, agreed on characteristics of particle size distribution by fractions are shown, approximately describing the inhalable fraction and subfractions which can enter the body through the larynx or up to nonciliated parts of the respiratory tract. The EN 481:1993 standard was introduced in Poland as the PN-EN:481:1998 standard. Particulate fraction capable of entering the human body through inhalation exposure is called **inhalable fraction** in the PN-EN:481:1998 standard, and the particulate fraction deposited in the gas exchange (respiratory) zone - the **respirable fraction**. The degree of dust penetration and deposition is dependent on particulate grain size. The inhalable fraction is composed of grains under the diameter of 100 μm. In the upper respiratory tract (nose, oral cavity, throat and larynx), large particles (over 30 μm in diameter) are retained and subsequently excreted with mucus. Particles with grain sizes under 7 μm can penetrate the gas exchange (respiratory) zone of the lung (the alveoli). The dust fraction capable of reaching the nonciliated parts of the respiratory tract is called the respirable fraction (see Fig. 1). It takes a long time to remove dust from the alveoli (approximately 50% a month). Mean aerodynamic diameter of a respirable fraction particle is 3.5 μm.

According to the authors [5] (Szewczyńska M., Pośniak M., 2017), it is necessary to replace probes used so far to determine total dust with probes ensuring quantitative separation of wood dusts respirable fraction in accordance with this fraction convention defined in the PN-EN 481:1998 standard. In the aforementioned article, results of studies conducted on workstations in seven furniture industry establishments were presented. Total dust and respirable fraction were determined using the gravimetric method in accordance with PN-Z-04030-05:1991 [6]

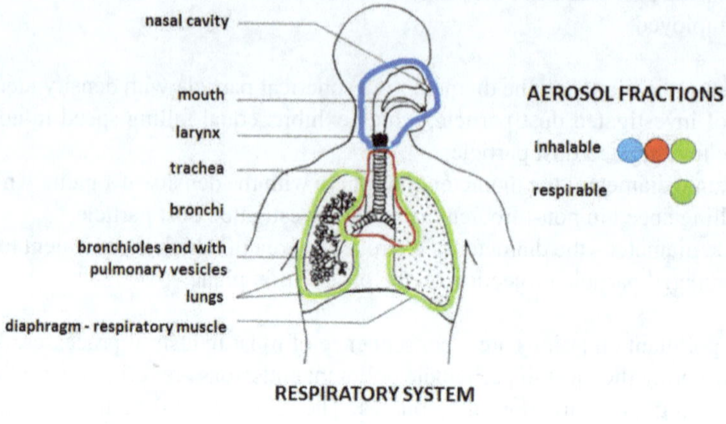

Figure 1: Respiratory system with aerosol fractions.

and PN-Z-04030-06:1991 [7] standards. Air samples for study purposes were collected using the individual dosimetry method. Total dust concentrations were established at 0.84–13.92 mg/m^3 and inhalable fraction concentrations, obtained from total dust concentrations (after applying a 1.59 coefficient) were 1.34–22.13 mg/m^3. Respirable fraction concentrations were 0.38–4.04 mg/m^3, which amounts to approx. 25% of inhalable fraction on average. The results of mixed wood dusts concentration measurements in furniture manufacturing establishments indicate a significant respirable fraction percentage in the inhalable fraction of such dusts.

Methods described in the withdrawn PN-91/Z-04030/05 and PN-91/Z-04030/06 standards are currently applied in Poland. These standards describe outdated methods for dust fractions determination, therefore it its necessary to establish new standards encompassing methods of inhalable and respirable fractions in compliance with the PN-EN 481:1998 standard recommendations. The main purpose of the study was the evaluation of chosen methods as tools for determination of the inhalable and respirable factions in dusts present in working conditions. In the future, they will serve the preparation of documentations for the draft projects of two Polish standards concerning the determination of the inhalable and respirable dust fractions.

2 METHODS AND INSTRUMENTS

2.1 Methodology for determining the inhalable and respirable fraction of industrial dusts

The most reliable results of occupational exposure assessment are obtained using an individual dosimetry method. The research was carried out at seven selected workstations where mostly wood dust or coal dust were emitted. The methodology of this research was based on studies described by [8, 9]. Due to fact, that side-by-side testing method was considered as the most practical, an approach based on this method was adopted in present study. Dust samples were collected in a breathing zone on a specially constructed stand containing a set of 8 or 14 sampling heads (4 pair of samplers for inhalable fraction, 3 pair of samplers for respirable fraction measurement) along with individual aspirators (see Fig. 2). At each of the

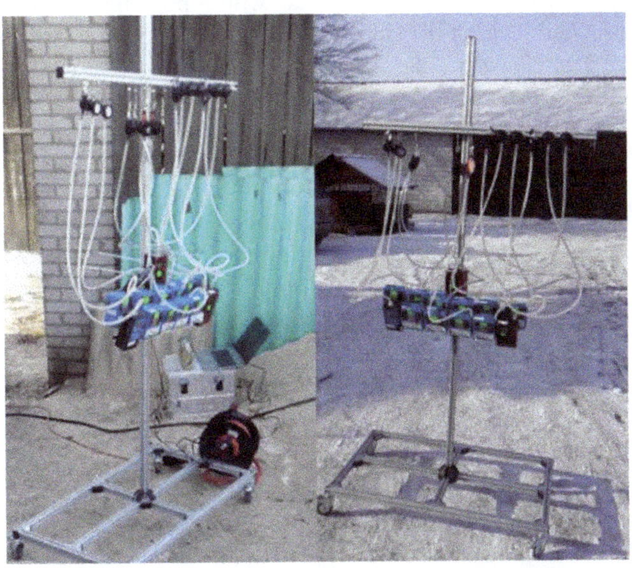

Figure 2: Arrangement of measuring sets on a stand.

investigated workstations, samples of the inhalable fraction and respirable fraction of the dust were collected simultaneously using different sets of sampling heads capable of sampling with varying volume air flow. For each studied sampling head an IOM sampler was used as a reference. The diagram of the IOM sampling head construction is shown in Fig. 3. This sampler was developed by J.H. Vincent and D. Mark in the IOM - Institute of Occupational Medicine [10–13]. Results of dust collection were referenced to an IOM sampler which value was set as 100% and then deviations were determined.

The gravimetric method was applied in order to determine inhalable and respirable dust fractions. Inhalable fraction determination procedure was carried out through dusty air suction with a known flowrate within a specified time through a measuring filter. Dust concentration was then established as a ratio of dust mass deposited on the measuring filter to the volume of suctioned air (see eqn. 1). The respirable fraction determination was carried out through dusty air suction with a known flowrate within a specified time through a microcyclone (a preselector, designed to retain large grain diameter fractions) and subsequently a measuring filter, on which the respirable dust fraction was deposited. Respirable fraction concentration was determined as a ratio of dust mass deposited on the measuring filter to the total volume of filtered air (see eqn. 1).

$$C = \frac{m_2 - m_1}{V} \times 1000 \tag{1}$$

where:

- m_2 – filter mass after sampling in milligrams,
- m_1 – filter mass before sampling in milligrams,
- V – air sample volume, calculated as a product of air flowrate and sampling time in dm^3.

As it is extremely important to keep the flowrate constant during air sampling, the air flowrates were checked before and after the process with a portable bubble flowmeter Gilian Gilibrator-2 (Sensidyne). During the study, environmental conditions which can potentially strongly influence dust concentrations, such as air temperature and relative humidity, were

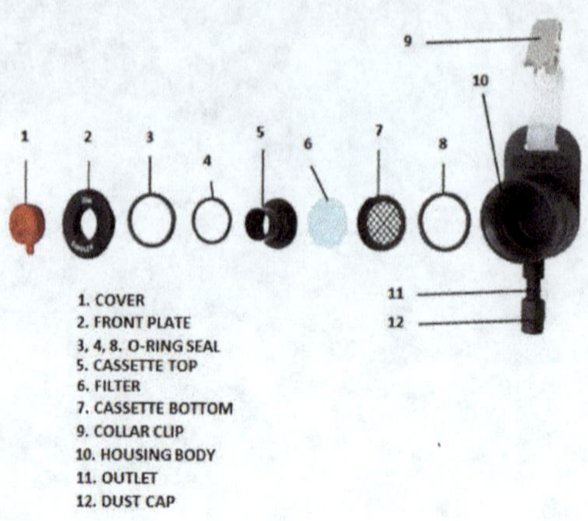

Figure 3: Diagram of the IOM sampling head construction.

closely monitored. Before the testing commenced, on each workstation data necessary for proper results interpretation were gathered, such as:

- the kind and nature of the work process in question, as well as the kind of dust-emitting materials used,
- work timing of particular employees and the total number of employees,
- applied individual and group protection measures.

A comparative analysis of test results for different probes used for inhalable and respirable dust fraction determination was carried out in the conditions of varying flow rates, depending on the work stations where dusts of different concentrations were emitted.

The PN-91-0430/05 standard allows for sampling of inhalable dust (inhalable fraction) through the use of a stationary or a personal dust meter. Regardless of the chosen sampling method, the mass of deposited dust should be no less than 0.5 mg and the total mass of dust collected on the filter no higher than 0.5 mg/cm^2.

The measurements of respirable dust (respirable fraction) can be carried out through the use of a personal or a stationary dust meter as per the PN-91 Z-04030/06 standard. Sample mass deposited on a filter should not exceed 5 mg. Sample collection time must allow for the deposition of at least a minimum allowed dust mass on the filter:

- 0.3 mg, if a scale with a 0.01 mg reading interval is used,
- 1 mg, if a scale with a 0.05 mg reading interval is used.

Inhalable fraction and respirable fraction exposure indicators (C_w) at the workstation were calculated as weighted averages for a working shift, in milligrams per cubic meter, using the following formula (2):

$$C_w = \frac{C_1 \times t_1 + C_2 \times t_2 + ... + C_n \times t_n}{t_1 + t_2 + ... + t_n} \tag{2}$$

where:

- C_1, C_2, ...C_n – concentrations obtained through sample determination, in milligrams per cubic meter,
- $t_1, t_2, ...t_n$ – time of sampling for a particular sample, in hours,
- n – number of samples.

The study was conducted at seven different workstations: 1. Sawmill I (hard wood dust - Oak), 2. Joinery I (hard wood dust: Oak, Ash, Beech), 3. Joinery II (wood dust: MDF - mixture of hard wood), 4. Sawmill II (dust of soft fresh coniferous wood), 5. Joinery III (wood dust: MDF - mixture of soft wood), Coal-Fired Power Plant (coal dust and biomass), Car Workshop (dust: mixture of calcium sulphate / gypsum with magnesium). As wood dust was sampled at five of the workstations, respirable fraction tests were not carried out due to the wood dust particles being too large (and due to lack of maximum admissible concentration for wood dust).

2.2 Measuring equipment

The selection of studied samplers for assessment and reference sampler was dictated by the equipment available on the Polish market and the results of the works and recommendations

described in [8–12, 14–18]. The following heads were used to determine the inhalable fraction: TWO-MET (open head), open head sampler EKOHIGIENA with an internal filter cassette (identical flow rate and construction highly like the TWO-MET probe), GSP 3.5 and GSP 10. The following heads were used to determine the respirable fraction: TWO-MET (cyclone separator), Higins-Dewell (with an upper inlet) and FSP 10. For each of the heads a reference IOM head was used, with an appropriately selected foam in case of the respirable fraction tests. Sampling heads are presented in Table 1.
The following equipment was used for sampling:

- **personal aspirators**, ensuring a steady airflow with a deviation no higher than 7% and flow rate ensuring proper operation of the measuring heads: Gilian-5000 (Sensidyne), GilAir-5 (Sensidyne), SG10-2 (GSA Messgerätebau GmbH),

Table 1: Sampling heads used for measurements.

Fraction	Name of the sampling head	Image	Air volume flow [dm^3/min]	Filter for inhalable / respirable fractions	Size of filter [mm]
Inhalable	1 – IOM		2.0	Glass / Polypropylene	25
	2 – TWO-MET with open head		2.0	Glass / Polypropylene	37
	3 – IOM		2.0	Glass / Polypropylene	25
	4 – EKOHIGIENA with open head		2.0	Glass / Polypropylene	37
	5 – IOM		2.0	Glass / Polypropylene	25
	6 – GSP 3.5		3.5	Glass / Polypropylene	37
	7 – IOM		2.0	Glass / Polypropylene	25
	8 – GSP 10		10.0	Glass / Polypropylene	37
Respirable	9 – IOM with foam for respiratory fraction		2.0	- / Polypropylene	25
	10 – TWO-MET cyclone separator		1.9	- / Polypropylene	37
	11 – IOM with foam for respiratory fraction		2.0	- / Polypropylene	25
	12 – Higgins Dewell with the upper inlet		2.2	- / Polypropylene	37
	13 – IOM with foam for respiratory fraction		2.0	- / Polypropylene	25
	14 – FSP 10	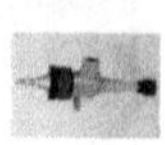	10.0	- / Polypropylene	37

- **measuring heads**: IOM - reference head (flow rate 2.0 dm^3/min), TWO-MET - head with one slot for a filter cassette (dedicated for welding fumes, flow rate 2,0 dm^3/min), EKO-HIGIENA - open 37 mm diameter head with inner filter cassette (flow rate 2.0 dm^3/min), GSP 3.5 (flow rate 3.5 dm^3/min), GSP 10 (flow rate 10.0 dm^3/min), to mount the frame where an employee breathing zone filter is placed.
- **cyclone selectors** for the purpose of respirable and inhalable fraction separation: IOM – reference head (flow rate 2.0 dm^3/min), Higgins-Dewell (flow rate 2.2 dm^3/min), TWO-MET head with a cyclone separator (flow rate 1.9 dm^3/min), FSP 10 (flow rate 10.0 dm^3/min). The cyclone selector is a vital part of the measuring system. Its flow rate must be adjusted in such a way to ensure respirable fraction intake. Otherwise, an intermediate phase will be deposited on the filter. The measurements will then apply to an entirely different fraction and the conclusions will be in error.
- **measuring filters** (non-hygroscopic filters with a filtration efficiency over 95%): polypropylene, glass, cellulose,
- **analytical scale** XPE with an interval no less than 0.01 mg (Mettler Toledo),
- **desiccator** for filter and humidity sensitive sample storage (WSL Sp. z o.o.),
- **thermo-hygro-barometer** LB-706 (with a LB-701 probe, LAB-EL),
- **bubble flow meter** Gilibrator-2 (Sensidyne).

3 RESULTS AND DISCUSSION

These studies were conducted to compare the various chosen samplers currently available on the Polish market. Results were taken from seven workstations: 1 - Sawmill I (hard wood dust - Oak), 2 – Joinery I (hard wood dust: Oak, Ash, Beech), 3 - Joinery II (wood dust: MDF - mixture of hard wood), 4 – Sawmill II (dust of soft fresh coniferous wood), 5 - Joinery III (wood dust: MDF - mixture of soft wood), 6 - Coal-Fired Power Plant (coal dust and biomass), 7 - Car Workshop (dust: mixture of calcium sulphate / gypsum with magnesium).

3.1 Pumps performance and environmental conditions

To keep the flowrate constant during air sampling, the air flowrates were checked before and after the process. Average volume flow of the pumps were measured from values checked before and after the process. Environmental conditions, which can potentially strongly influence dust concentrations, were checked each hour of the measurements and average value was calculated (see Table 2).

All flows rates were maintained at a good level for dust sampling. Exception was the flow recorded by the pump connected to EKOHIGIENA sampler at the second workstations where the flow rate was 8.57% (should not exceed 5%). However, the amount of dust collected by this sampler did not differ significantly from the reference value.

3.2 Mass concentrations of collected dust

Mass concentrations of dust collected at seven different workstations are presented in Table 3. By analysing the results, it was found that the IOM sampler at each workstation obtained similar values and were good choice as a point of reference to the other sampling heads. Small discrepancies between IOM samplers were probably caused by the distance between the measuring pairs arranged on a stand. In three cases, no dust or only trace amounts of dust were collected on the filters (results rejected). As checks of the pumps have shown, the air

Table 2: Average volume flow for the pumps and environmental conditions during measurements.

Type of head connected to a specific pump	Average volume flow [dm³/min] (% change of indication)						
	Workstation						
	1	2	3	4	5	6	7
1 – IOM ref.	2.009 (1.64%)	2.003 (0.31%)	2.015 (0.39%)	1.995 (0.21%)	1.997 (0.74%)	2.000 (0.16%)	2.026 (2.63%)
2 – TWO-MET with open head	2.043 (4.57%)	2.034 (3.49%)	2.010 (0.38%)	1.998 (1.14%)	2.000 (0.38%)	2.012 (1.40%)	2.001 (0.09%)
3 – IOM ref.	2.017 (1.71%)	1.996 (1.11%)	1.993 (0.20%)	2.019 (0.51%)	2.003 (0.26%)	1.975 (2.09%)	2.021 (2.74%)
4 – EKOHIGIENA, open head with inner cassette	2.008 (0.85%)	2.089 (**8.57%**)	2.003 (0.60%)	1.997 (1.32%)	2.006 (0.54%)	1.995 (0.18%)	2.002 (0.37%)
5 – IOM ref.	2.022 (2.27%)	2.010 (0.84%)	2.032 (2.33%)	2.012 (1.62%)	2.021 (2.63%)	2.008 (0.51%)	1.999 (0.24%)
6 – GSP 3.5	3.536 (2.01%)	3.532 (1.16%)	3.540 (0.47%)	3.506 (0.10%)	3.499 (0.41%)	3.493 (0.09%)	3.525 (1.24%)
7 – IOM ref.	2.040 (3.81%)	1.994 (%1.05)	1.976 (1.62%)	1.995 (1.44%)	1.987 (1.34%)	2.003 (0.38%)	1.993 (0.83%)
8 – GSP 10	10.247 (4.29%)	10.012 (0.25%)	10.146 (2.51%)	10.003 (0.45%)	10.038 (0.91%)	10.072 (1.84%)	10.163 (2.11%)
9 - IOM ref.	-	-	-	-	-	2.005 (0.88%)	2.005 (0.40%)
10 - TWO-MET with cyclone separator	-	-	-	-	-	1.898 (0.26%)	1.909 (0.28%)
11 – IOM with foam for respiratory fraction	-	-	-	-	-	1.994 (0.37%)	2.000 (0.33%)
12 – Higgins Dewell with the upper inlet	-	-	-	-	-	2.194 (0.63%)	2.197 (0.10%)
13 – IOM with foam for respiratory fraction	-	-	-	-	-	1.994 (0.11%)	2.006 (1.01%)
14 – FSP 10	-	-	-	-	-	10.371 (0.83%)	10.256 (3.13%)
	1	2	3	4	5	6	7
Air temperature [°C]	-1.60 ÷ 1.90	23.7 ÷ 24.0	18.4 ÷ 24.5	16.0 ÷ 17.4	18.3 ÷ 18.5	25.6 ÷ 27.3	22.7 ÷ 24.9
Relative air humidity [%]	62.90 ÷ 72.10	**37.8** ÷ 38.2	**31.0** ÷ 35.7	66.6 ÷ 71.0	61.0 ÷ 63.0	31.5 ÷ 42.0	50.6 ÷ 61.5
Sampling time [min]	303	316	263	242	223	170	366

Table 3: Mass concentrations of collected dust.

		Mass concentration of dust [mg/m^3]						
		Workstation						
Head number	Fraction	1	2	3	4	5	6	7
1 (ref IOM)	Inhalable	1.16	0.38	0.49	0.75	2.35	4.29	9.47
2		0.97	0.30	0.40	0.13	1.48	4.68	9.95
3 (ref IOM)		1.38	0.38	0.41	0.46	2.00	5.27	10.02
4		1.29	0.50	0.62	0.25	1.25	11.15	9.78
5 (ref IOM)		1.95	0.34	0.46	0.54	1.68	5.31	9.75
6		1.93	**0.00**	**0.00**	2.33	3.62	2.90	11.35
7 (ref IOM)		1.63	0.67	0.47	0.79	2.02	5.60	9.94
8		2.18	0.23	0.30	0.88	0.62	3.86	9.88
9 (ref IOM)	Respirable	-	-	-	-	-	0.26	1.25
10		-	-	-	-	-	0.20	1.93
11 (ref IOM)		-	-	-	-	-	0.14	1.22
12		-	-	-	-	-	0.17	1.71
13 (ref IOM)		-	-	-	-	-	0.56	0.86
14		-	-	-	-	-	**0.00**	2.89

flow was undisturbed and they worked properly, also filters were not clogged by larger objects (like leaves, plastic pieces, paper scraps, etc.). Lack of dust on the filters could be caused by leaks, poorly placed or damaged filter. Interestingly, on the days when there was a problem with dust collection (workstation 2 and 3) humidity was relatively lower than the other days. At the second workstation, also the ECKOHIGIENA sampler pump had disturbed airflow. With so few results it is not a determinant, but it is worth keeping in mind for further research.

3.3 Deviation from the reference values

Results of dust collection were referenced to IOM samplers which value was set as 100% and then deviations were determined as percentage differences (see Fig. 4). Average percentage deviation from reference value shown that the biggest discrepancies can be seen with the GSP 3.5 sampler for inhalable fraction (over 100%). EKOHIGIENA sampler had discrepancies around 40%, GSP 10, around 35% and TWO-MET sampler (with open head) around 27%. Samplers: Higgins Dewell and TWO-MET (with cyclone separator) used for respirable fractions had discrepancies around 30–38% and FSP 10 collected nearly 2.5 times more dust than reference IOM sampler, but this is only based on one result. Second result for this sampler was rejected because no dust or only trace amounts of dust was collected. Same problem appeared twice in case of GSP 3.5 sampler.

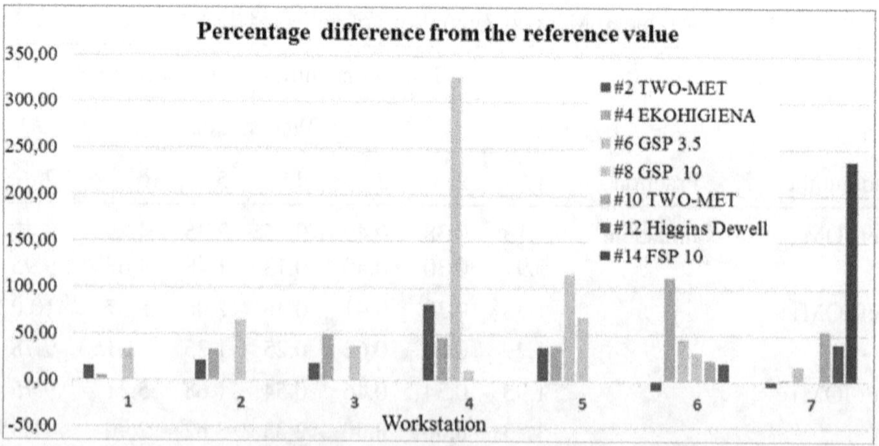

Figure 4: Percentage deviations from reference value.

3.4 Exposure indicators

Measurement of dust concentrations allowed to determine the level of exposure. Workplace activities and the type of dust the worker was exposed to: 1. Woodcutter - cutting boards (hard wood; oak dust), 2. Carpenter - cutting board (hard wood; oak, ash, beech dust), 3. Carpenter - cutting MDF (medium-density fibreboard, mixture of hard wood). 4. Woodcutter – cutting of fresh coniferous wood (soft coniferous wood dust), 5. Carpenter - cutting MDF (medium-density fibreboard, mixture of soft wood), 6. Physical worker - staying in a pouring building at the level of the sifters (coal dust and biomass), 7. Grinder - grinding of the filler compound (mixture of calcium sulphate with magnesium). See Table 4.

At the workstations where wood, coal and calcium sulphate dust was emitted no threshold values for inhalable and respirable fractions were exceeded. Only at the car wokshop values for inhalable fraction was slightly exceeded.

4 CONCLUSION

Pilot studies have shown that a method of simultaneously collected dust at workstations with use of reference samplers gives an opportunity to determine size fractions of an aerosol, exposure indicators and to evaluate selected samplers in view of new definitions in Polish regulations. Unfortunately, we did not avoid errors during the measurements, and the method itself needs to be modified. In subsequent studies the most important will be an increase of the measurement statistics. Next studies are planned, but with repeating the dust collection at the respective worksites for at least another two days (at least 3 samples for each sampling head). As the amounts of dusts present in work environments are getting progressively smaller, currently applied mass based approaches may potentially be not precise enough for the purpose of such study. This is the reason for which dust concentration measurements in further studies will be simultaneously carried out with use of standardized testing methods and measuring instruments capable of detecting mass concentration (mg/m^3) and particles number concentration (number of particles/cm^3) like portable aerosol spectrometers and ultrafine particle counters.

Table 4: Exposure indicators for different studied activities.

Head number	Workstation / activities; *Cw* [mg/m^3] ± expanded uncertainty						
	1	2	3	4	5	6	7
1	1.17 ± 0.12	0.38 ± 0.04	0.50 ± 0.05	0.74 ± 0.08	2.35 ± 0.08	4.29 ± 0.44	9.59 ± 0.98
2	0.94 ± 0.10	0.30 ± 0.03	0.40 ± 0.04	0.13 ± 0.01	1.48 ± 0.01	4.71 ± 0.48	9.96 ± 1.02
3	1.39 ± 0.14	0.38 ± 0.04	0.41 ± 0.04	0.46 ± 0.05	2.00 ± 0.05	5.20 ± 0.53	10.13 ± 1.03
4	1.30 ± 0.13	0.53 ± 0.05	0.62 ± 0.06	0.25 ± 0.03	1.25 ± 0.03	11.12 ± 1.13	9.79 ± 1.00
5	1.97 ± 0.20	0.34 ± 0.04	0.47 ± 0.05	0.55 ± 0.06	1.70 ± 0.06	5.34 ± 0.54	9.74 ± 0.99
6	1.94 ± 0.20	**0.00**	**0.00**	2.33 ± 0.24	3.62 ± 0.24	2.90 ± 0.30	11.44 ± 1.17
7	1.66 ± 0.17	0.67 ± 0.07	0.47 ± 0.05	0.78 ± 0.08	2.01 ± 0.08	5.61 ± 0.57	9.91 ± 1.01
8	2.23 ± 0.23	0.23 ± 0.02	0.30 ± 0.03	0.88 ± 0.09	0.63 ± 0.09	3.89 ± 0.40	10.04 ± 1.02
9						0.26 ± 0.03	1.25 ± 0.13
10						0.20 ± 0.02	1.94 ± 0.20
11						0.14 ± 0.01	1.22 ± 0.12
12						0.17 ± 0.02	1.70 ± 0.17
13						0.56 ± 0.06	0.86 ± 0.09
14						**0.00**	2.97 ± 0.3
Average	1.58 / -	0.40 / -	0,45 / -	0.77 / -	1.88 / -	5.38 / 0.27	10.08 / 1.66
Threshold limit Value [mg/m^3] Inhalable / respirable	2 / -	2 / -	2 / -	4 / -	4 / -	10 / 2	10 /2

ACKNOWLEDGEMENTS

This paper has been based on the results of a research task carried out within the statutory activity of Central Institute for Labour Protection – National Research Institute (task number I-52/TSB). Studies were conducted to compare the various samplers currently available on the Polish market. This article does not constitute an advertisement of any of the listed companies. The publication is of a review nature. Information were obtained from companies producing and distributing heads for respiratory and respiratory fractions (websites: SKC, Zeff, Casella, Ekohigiena).

REFERENCES

[1] Zgierska, A., Working Conditions in 2016, *Central Statistical Office Report* (pol. Główny Urząd Statystyczny, GUS), Warsaw 2016.

[2] Więcek, E., Health criteria for aerosol sampling in the work environment, *Principles and Methods of Assessing the Working Environment*, 2011.

[3] PN-EN 481:1998, Polish Standard for: Workplace atmospheres – Size fraction definitions for measurement of airborne particles, 1998.

[4] EN 481:1993, European Standard: Workplace atmospheres – Size fraction definitions for measurement of airborne particles, 1993.

[5] Szewczynska, M. & Posniak, M., Assessment of occupational exposure to wood dust in the Polish furniture industry. *Occupational Medicine*, **68**(1), pp. 45–60, 2017. https://doi.org/10.13075/mp.5893.00458

[6] PN-91/Z-04030/05, Polish Standard for: Air purity protection. Investigation of dust content. Determination of total dust at work stations by filter-weight method, 1991.

[7] PN-91/Z-04030/06, Polish Standard for: Air purity protection. Investigation of dust content. Determination of total dust at work stations by filter-weight method, 1991.

[8] Lidén, G, Juringe, L. & Gudmundsson, A., Workplace validation of a laboratory evaluation test of samplers for inhalable and “total” dust. *Journal of Aerosol Science*, **31**(2), pp. 199–219, 2000.
https://doi.org/10.1016/s0021-8502(99)00049-x

[9] Lee, T., Harper, M., Slaven, J.E., Lee, K., Rando, R.J. & Maples, E.H., Wood dust sampling: field evaluation of personal samplers when large particles are present. *Annals of Occupational Hygiene*, **55**(2), pp. 180–191, 2011.
https://doi.org/10.1093/annhyg/meq075

[10] General methods for sampling and gravimetric analysis of respirable and inhalable dust. *HSE, MDHS14/3,* 2000.

[11] Sanchez Jimenez, A., van Tongeren, M. & Aitken, R.J., Guidance for collection of inhalable and respirable in dust, Strategic Consulting. *Institute of Occupational Medicine*, 538-0000, 2012.

[12] Sanchez Jimenez, A., van Tongeren, M. & Cherrie, J.W., A review of a monitoring methods for inhalable hardwood dust, Research Report. *Institute of Occupational Medicine,* P937/1A, 2011.

[13] LEE, T., Kim, S.W., Chisholm, W.P., Slaven, J. & Harper, M., Performance of high flow rate samplers for respirable particle collection. *Annals of Occupational Hygiene*, **54**(6), pp. 697–709, 2010.
https://doi.org/10.1093/annhyg/meq050

[14] Kenny, L.C., Aitken, R., Chalmers, C., Fabries, J., Gonzalezfernandez, E., Kromhout, H., Liden, G., Mark, D., Riediger, G. & Prodi, V., A collaborative European study of personal inhalable aerosol sampler performance. *Annals of Occupational Hygiene*, **41**(2), pp. 135–153, 1997.
https://doi.org/10.1016/s0003-4878(96)00034-8

[15] Li, S.-N., Lundgren. D.A. & Rovell-Rixx, D., Evaluation of six inhalable aerosol samplers. *AIHAJ A Journal for the Science of Occupational and Environmental Health and Safety,* **61**(4), pp. 506–516, 2000.
https://doi.org/10.1080/15298660008984562

[16] Kenny, L., Chung, K., Dilworth, M., Hammond, C., Jones, J.W., Shreeve, Z. & Winton, J., Applications of low-cost, dual-fraction dust samplers. *Annals of Occupational Hygiene*, **45**(1), pp. 35–42, 2001.
https://doi.org/10.1016/S0003-4878(00)00008-9

[17] Görner, P., Simon, X., Wrobel, R., Kauffer, E. & Witschger, O., Laboratory study of selected personal inhalable aerosol samplers. *Annals of Occupational Hygiene*, **54**(2), pp. 165–187, 2010.
https://doi.org/10.1093/annhyg/mep079

[18] Verpaele, S. & Jouret, J., A comparison of the performance of samplers for respirable dust in workplaces and laboratory analysis for respirable quartz. *Annals of Occupational Hygiene*, **57**(1), pp. 54–62, 2013.
https://doi.org/10.1093/annhyg/mes038

Author index

www.ingramcontent.com/pod-product-compliance
Lightning Source LLC
LaVergne TN
LVHW062238200926
848484LV00006B/1571

9781784663131